Learning to Fly Helicopters

TAB
PRACTICAL
FLYING SERIES

Other Books in the TAB PRACTICAL FLYING SERIES

Learning to Fly Helicopters

R. Randall Padfield

Illustrated by
Ralph C. Padfield

TAB Books
Division of McGraw-Hill

New York San Francisco Washington, D.C. Auckland Bogotá
Caracas Lisbon London Madrid Mexico City Milan
Montreal New Delhi San Juan Singapore
Sydney Tokyo Toronto

© 1992 by **TAB Books**.
TAB Books is a division of McGraw-Hill, Inc.

pbk   9 10 11 12 13 FGR/FGR 9 9
hc   1 2 3 4 5   6   7   8   9 FGR/FGR 9 9 8 7 6 5 4 3 2

Library of Congress Cataloging-in-Publication Data

Padfield, R. Randall.
    Learning to fly helicopters / by R. Randall Padfield.
        p.   cm.
    Includes index.
    ISBN 0-8306-2113-X (hardback)   ISBN 0-8306-2092-3 (pbk.)
    1. Helicopters—Piloting.   I. Title.
TL716.5.P34   1992   91-26770
629.132′5252—dc20   CIP

Acquisitions Editor: Jeff Worsinger
Book Editor: Norval G. Kennedy
Director of Production: Katherine G. Brown
Book Design: Jaclyn J. Boone   PFS
Series Design: Jaclyn J. Boone   3763

Contents

12 Human Factors and Safety   237

13 Born-again Copilots   257

14 Ten Commandments   265

Foreword

WHEN I RECEIVE A PHONE CALL OR LETTER FROM SOMEONE OTHER THAN a professional in the rotary-wing industry, it is almost always from someone wanting information on how to fly helicopters. Young men and women seeking a career, people desiring a career change, fixed-wing pilots who want to augment their aviation skills—the callers and letter writers hold various motivations, but share a single, often fervent, desire: learning to fly an aircraft that hovers.

Thanks to R. Randall "Randy" Padfield, I can now refer these people to a commendable source. Randy writes about the often-complex subject of helicopter aerodynamics and flying in a clear, understandable style. Indeed, his style is one with which I've grown quite familiar, because he has regularly written feature articles for *Rotor & Wing International* and other aviation publications.

There are other books on how to fly helicopters, but Randy's offers some unique, and I believe quite important, features. One is "Helicopter Myths." Even among seasoned fixed-wing pilots, there are misconceptions regarding helicopters, and many of these restrain in part rotorcraft's advancement. It's refreshing to see these misconceptions rebutted squarely.

Another important feature is the chapter covering emergencies. Though infrequent, emergencies do happen; however, Randy informs readers on how they can develop what he calls the "right stuff" and stay clear of harm's way. As Randy points out, this involves a proper attitude as much as proficient piloting skills.

But probably most important, Randy's writing can be distinguished from that in other how-to-fly-a-helicopter books because it is fun to read. And appropriately so, for I'm convinced fun represents much of the reason many of the callers and letter writers I encounter want to learn to fly helicopters.

David Jensen, Executive Editor
Rotor & Wing International

To my wife, Moira,
and our children,
Heitha, Dirk, and Thomas

Acknowledgments

I WANT TO THANK MY PARENTS, RALPH AND CLARA ANNE PADFIELD, for their help in the preparation of *Learning to Fly Helicopters*.

With great care and concern for technical accuracy, my father prepared the numerous drawings found throughout the book. He also read the manuscript, checking it for clarity and accuracy. His experience as a Navy pilot during World War II and his years as an engineer and manager for Bethlehem Steel Research provided him with an aviation and technical writing background that made his comments and suggestions both knowledgeable and succinct.

As a very competent teacher and tutor, my mother was the perfect nontechnical reader for the book. When the original draft became overburdened with mechanical technicalities, she noticed it; when the grammar was not quite what it should be, she corrected it; when I belabored a point (a characteristic I get from my father), she pointed it out. Thanks to my parents, the book you now hold in your hands is clearer, more accurate, and easier to read.

I also want to thank all the helicopter manufacturers that provided photographs and information about their aircraft: Aerospatiale Helicopter Company, Agusta Aerospace Corporation, Bell Helicopter Textron, Boeing Helicopters, E.H. Industries, Enstrom Helicopter Corporation, MBB Helicopter Corporation, McDonnell Douglas Helicopter Company, Robinson Helicopter Company, Rogerson Hiller Corporation, Schweizer Aircraft Corporation, United Technologies Sikorsky Aircraft, and Westland Incorporated. Thanks, too, to FlightSafety International for photographs of their helicopter simulators.

Finally, special thanks to David Jensen, executive editor of *Rotor & Wing International* Magazine, for the foreword; to J. Mac McClellan, editor-in-chief of FLYING Magazine, for allowing me to use portions of his article, "Pilot, Superpilot," in chapter 6; and to Bell Helicopter Textron for allowing me to use portions of their training guide, "Flying Your Bell Model 206 JetRanger," in chapter 7.

Introduction

WHAT'S IT LIKE TO FLY A HELICOPTER? IT'S A FEELING. A FEELING THAT must be experienced to be understood. The best I can do is identify other experiences that create similar feelings, experiences you're more likely to have had, and hope you'll get the idea.

It's a feeling you get when playing football. You go out for a long pass. You're running full speed down the field and hear your name called out. You look over your shoulder and there's the ball—right there—floating toward you. You reach out in front of you, your legs pumping as fast as they can go, and the ball just settles right into your fingers as if it were weightless. You grasp the ball, cradle it to your side, and feel like you could run on forever.

Or you're playing baseball. The pitcher throws and before the ball is halfway to the plate you know it's the kind of pitch you like. You time it just right and swing hard and smooth. The bat meets the ball full on, a loud satisfying "thunk." As the bat coils around your body, you watch the ball sail straight out over the pitcher's head in a perfect 45-degree arc and it sails on and on and on.

It's a feeling, a hands-eyes-and-feet coordination thing. The kind of feeling a gymnast gets on the balance beam. The kind of feeling an airplane pilot gets when he squeaks on a landing in a taildragger. The kind of feeling a skier gets coming down a slope covered in new powder. The kind of feeling everyone gets when they learn to ride a bicycle without training wheels.

That's what it's like to fly a helicopter.

You won't get the feeling riding as a passenger in a helicopter. You won't get it the first time you fly a helicopter or even the second or the third. When you're learning, you'll be concentrating too much on the basics to get the feeling. You have to master the basics before the feeling comes. Be patient. It will come with time.

When you get the feeling, you'll know it. The helicopter will no longer feel like an alien machine trying to kill you at every turn. You won't think of it as thousands of individual parts flying together in loose formation while they try to beat the air into submission. You won't think of it as the "inherently unstable" ugly duckling of the aviation industry. When you get the feeling, the helicopter will become your magic carpet.

This book can't give you the feeling. No book can. What this book can do is pave the way so that you'll get the feeling sooner.

Helicopters are fascinating, complicated machines. They're not easy to fly and they're not easy to understand. You have a lot of study and work and practice ahead of you if you are to become a helicopter pilot. Believe me, it's worth the time, effort, and money.

Once you get the feeling, you'll never want to let it go.

1
Helicopter Myths

If God had wanted man to fly, he would have given him O.D. fire-resistant skin and pockets with zippers.

Unknown United States Army helicopter pilot, referring to the olive drab-colored Nomex flight suits worn by military pilots

ADMIT IT. DEEP DOWN ONE THING YOU'VE ALWAYS WANTED TO DO IS FLY a helicopter. Ever since you saw your first helicopter hovering over the ground, you've wondered what it's like to be a real "hover lover." But something has always held you back.

Maybe it's that number one horror story about helicopters: The engine stops and down you go, with all the glide ratio of a brick. Even twin-engine helicopters aren't safe, you've heard. And what about strong winds? Aren't they a problem for those fragile-looking whirlybirds? Most people will tell you no one in their right mind would really want to fly such unsafe aircraft. Why, you'd be risking your life every time you went up.

Hold on a minute. Let's clear up these things right from the start. First of all, let me assure you that all the horror stories about helicopters are just that—stories. The truth is helicopters and their pilots suffer from an image problem. Because of this and a general lack of knowledge about rotary-wing aircraft, a number of misunderstandings about helicopters, myths, if you will, have grown up over the years.

I've talked with many people—passengers, nonpassengers, even experienced air-plane pilots—and I've found that a few subjects are brought up time and time again (FIG. 1-1). Let's look at them one at a time.

Myth #1: If the engine quits, you're a goner.

Myth #2: Helicopters need two engines: one for the big propeller on the top and one for the little propeller in the back.

Myth #3: Helicopters are too fragile to fly in strong winds.

Myth #4: A flight in a helicopter is always bumpier than a flight in an airplane.

Myth #5: Helicopter pilots are different from other people.

Fig. 1-1. *The five myths about helicopters.*

MYTH #1:
If the engine quits, you're a goner

The film and television industries perpetuate this myth by constantly showing hel-icopters spinning madly out of control whenever the pilot so much as scratches his nose . . . not to mention when the movie villain does something mysterious, but obvi-ously foul, to the hero's machine. For an apprehensive viewer with little or no mechanical knowledge, it's easy to believe that it doesn't take much to make a helicop-ter fall out of the sky.

On the other hand, some people with limited aeronautical knowledge, even many fixed-wing pilots, hold fast to this myth. They reason that rotary-wing aircraft have glide ratios not much better than bricks or anvils. Therefore, when its engine stops, a single-engine helicopter is doomed to descend at such a high rate that a crash is inevi-table.

An object's glide ratio is the relationship between the distance it will travel unpowered over the ground compared to the height that it started gliding from; gliders are made to glide and therefore have good glide ratios; small airplanes usu-ally have fair glide ratios and supersonic jet aircraft have relatively poor glide ratios; bricks, anvils, and rocks obviously don't glide very far so they have extremely poor glide ratios (FIG. 1-2).

Helicopters don't have the best of glide ratios, but as long as the rotor blades keep turning, helicopters can do something airplanes can't do. And it's even better than gliding. It's called *autorotation*.

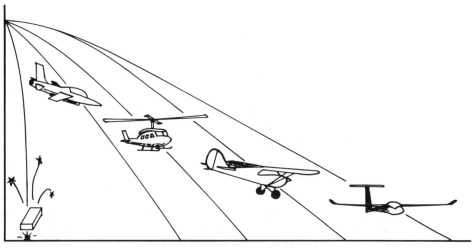

Glide distance (not proportional)

Fig. 1-2. *Relative glide ratios of several objects. A helicopter in autorotation has a better glide ratio than a supersonic jet aircraft.*

The fact is: You have a better chance of survival after a complete power failure in a helicopter than you do after a complete power failure in an airplane.

Helicopters can autorotate because they have rotating wings (rotor blades) instead of fixed wings. Think of the rotor blades on top of a helicopter as a fan. When you switch on a fan, an electric motor turns the fan's blades and the blades create a small breeze.

The opposite of a fan is a windmill. A windmill uses breezes and winds to drive pumps, generators, or other machinery. Air moves the blades of a windmill to drive machinery, whereas, the motor in a fan turns its blades in order to move the air.

A helicopter can act like either a fan or a windmill.

Most of the time, a helicopter acts like a fan. The engine turns the blades, the blades create lift, and the aircraft flies. But if the engine stops, the air flowing past the rotor blades, the *relative wind*, causes the blades to turn like a windmill. This allows a helicopter to make a controlled descent and landing.

What happens when the engine fails in a single-engine helicopter? (We'll get to twin-engine helicopters in Myth #2.)

The first event is the immediate and automatic disconnection of the engine from the rotor system by a freewheeling unit in the main transmission (FIG. 1-3). The effect is similar to when you stop pedalling a bicycle when going downhill. Because of your momentum and the pull of gravity, the bicycle's wheels continue to turn even though the "engine" has stopped. You might even pick up speed as you coast down the hill.

A flying helicopter is also subject to the force of gravity and it will continue "downhill" with its rotor blades "coasting" because of the effect of the relative wind turning them like a big windmill.

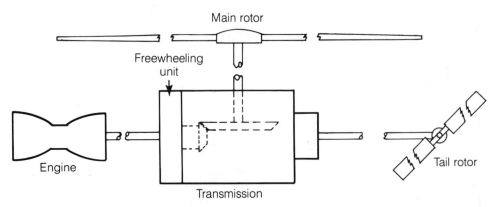

Fig. 1-3. *Simple schematic of a single-engine helicopter. The engine is coupled to the transmission via a freewheeling unit. In the event of an engine failure, the freewheeling unit automatically disconnects the engine from the transmission so that the main and tail rotors are free to autorotate.*

The net result is that helicopters do not glide like bricks, they do not fall from the sky like anvils, and they do not spin around like whirling dervishes when the engine fails. What they do is autorotate.

Although a helicopter in autorotation will descend at a faster-than-normal rate, helicopter pilots are trained to handle this event. As the helicopter nears the ground, the pilot manipulates the controls so that the momentum generated by the turning rotors during the descent is converted into lift. Some helicopters have so much energy that they can actually hover over the ground for a few seconds at the bottom of the autorotation.

The amount of lift available is dependent upon the weight of the helicopter, the temperature, the air pressure, and the surface wind. However, even under the most unfavorable conditions, a skilled pilot can still make a safe autorotative landing— no damage and no injuries—into an area not much larger than the helicopter itself (FIG. 1-4).

The ability to safely land in a small area is the main advantage an unpowered helicopter has over an unpowered airplane, but there are other advantages, too.

When an airplane loses all engine power, its electrical generators and hydraulic pumps stop, as well; however, because the generators and pumps in a helicopter are connected to the main transmission, as long as the rotor blades are turning, so are the generators and pumps.

This means that the helicopter pilot can use the same equipment during autorotation that he has available in powered flight: radios, navigation aids, autopilot system, and the like. An unpowered airplane, on the other hand, would be reduced to battery power alone, which usually means that some electrical consumers are lost. Small airplanes do not have hydraulically boosted controls, but the loss of total hydraulic power in a large airplane is a serious emergency. This would happen if an airplane had a total engine failure.

Fig. 1-4. *At the hands of a skilled pilot, a helicopter can make an engine-out landing into an area not much larger than the helicopter: Schweizer 300.*

So, you can see that autorotation is a very handy thing for the helicopter pilot to have.

MYTH #2:
Helicopters need two engines—
one for the big propeller on the top and
one for the little propeller in the back

Can you figure out one of the fallacies in this statement from the preceding explanation?

Think of a single-engine helicopter. It has a main rotor on the top, the "big propeller" (but don't ever call it that), and a tail rotor in the back, the "little propeller" (ditto), and it has but one engine; therefore, something else besides a second engine must make the little propeller—excuse me, the tail rotor—in the back go around.

That something is the same in both single- and twin-engine helicopters, even three-engine helicopters (yes, there are some, the EH 101, for example).

For comparison, a car has an engine. The engine turns the gears in the transmission and the transmission transfers the power to the wheels. In a normal two-wheel drive car, there is one engine powering two wheels.

What if you decided you wanted a more powerful car? You could, of course, take out the engine and install a bigger one. But, for the sake of this analogy, let's say that you decide to add another engine and connect it directly to the transmission.

Now you would have a car with two engines powering two wheels through a single transmission. If one engine were to stop, you could continue tooling on down the highway because you would still have power to both wheels from the engine that's still working.

A twin-engine helicopter is similar to that twin-engine car, except that the transmission of the helicopter drives the main rotor and the tail rotor, instead of two wheels (FIG. 1-5). Each engine has a freewheeling unit so that if one engine fails, it will not slow down the transmission and make it harder for the other engine to keep the rotors turning.

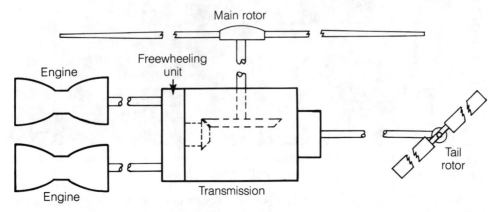

Fig. 1-5. *Simple schematic of a twin-engine helicopter. The transmission is powered by two engines that each have their own freewheeling gear. If one engine fails, the other engine can still provide power to main and tail rotors via the transmission.*

The reason a "standard" helicopter has a tail rotor is to counteract the torque of the main rotor. Without an anti-torque device to counteract the rotation of the main rotor, the fuselage of the helicopter would rotate in the opposite direction. Other ways of counteracting torque include the tandem rotors of the Boeing 234 Chinook or blowing pressurized air out vents in the tailboom like the McDonnell Douglas NOTAR (NO TAil Rotor), but we won't get into them just yet.

Why two engines? The obvious reason is to increase safety. Even though aircraft engines rarely fail, they can theoretically stop at any time, and the ability to continue flight on the remaining engine gives the pilot of a twin-engine helicopter more options (FIG. 1-6). The pilot of a single-engine helicopter has only one option available if the engine fails: autorotation. As discussed earlier, this is a very good thing to have, but it does mean the flight will end sooner than planned.

Numerous minor things can plague engines: partial failures of the control mechanism, stuck throttles, hiccups in the fuel system, and environmental factors, such as icing, heavy rain, and salt water spray, which although not always serious, can be cause for concern. If plagued by one of these problems, the pilot of a single-engine helicopter might consider it prudent to make an immediate precautionary landing.

Bell Helicopter Textron

Fig. 1-6. *If one engine fails in a twin-engine helicopter, the flight can be continued to the nearest safe landing area: Bell 222.*

Meanwhile, the pilot of a twin-engine helicopter, although concerned, might be able to continue the flight to the destination.

The fact is: If you were flying as a passenger in a twin-engine helicopter and one of the engines failed, you probably wouldn't even notice it.

More than once, I've had engine problems with one engine while piloting a twin-engine helicopter. Although the engines did not stop completely, they only provided a portion of their available power. To regain full power, I simply used the emergency fuel control lever in the cockpit to take manual control of the engine and continued to the destination.

The point is: None of the passengers on these flights ever realized we had any problem at all. We didn't lose altitude and airspeed decreased only slightly. After I brought in manual control, airspeed went back to normal. Absolutely no unusual noises, vibrations, movements, or other indications were noticed. We just kept on flying as if nothing had happened, and for all intents and purposes, nothing had happened. The other pilot and I only needed to give the malfunctioning engine a bit more attention than usual.

MYTH #3:
Helicopters are too fragile
to fly in strong winds

What's a strong wind? 20 knots? 40 knots? 60 knots? 100 knots?

What's a knot?

Okay, first things first; *knot* means nautical miles per hour. If you've ever done any boating, you're probably familiar with the use of knots as a measurement of speed.

Aviators chose to use knots to measure speed because sailors use knots to measure speed. (Sailors also tie knots, but that's another subject.) Sailors use knots because the world is divided into degrees of latitude and longitude. Degrees of latitude and longitude are divided into minutes, 60 minutes equalling one degree. One nautical mile equals one minute of arc on a meridian (one minute of latitude) or one minute of arc on the earth's equator. All nautical and aeronautical charts are marked off in degrees of latitude and longitude; therefore, it only makes sense to use nautical miles per hour when using these charts. If you were to use statute miles per hour or kilometers per hour with such charts, a conversion factor would constantly have to be applied.

Back to our original question: What are strong winds?

If you stand in the middle of an open field and get hit by 40 knots of wind, you'd probably agree that it feels like a strong wind. Sixty knots will push you over, if you don't lean into it, and a 100-knot wind will make you crawl. (By the way, meteorologists classify any sustained winds in excess of 64 knots as hurricane force.)

Let's say that anything above 40 knots is a strong wind to someone standing on the ground. How does 40 knots of wind feel to an aircraft? Like nothing at all.

Wind has no meaning to an object in flight, except in its relationship to the ground. Once an aircraft—this applies to airplanes, gliders, helicopters, balloons, gyroplanes, and the like—rises above the ground, it becomes one with the wind. The ground could disappear, for that matter, and it would make no difference to an aircraft (FIG. 1-7). If the ground did disappear, flight at lower altitudes would be smoother. Much of the turbulence at lower altitudes is caused by the uneven heating of the earth's surface or by movement of the air over and around the terrain.

A balloon, for example, can only move with the air mass; therefore, with the wind. If you lay a sheet of paper on a table in the gondola under a balloon, the paper will remain completely motionless, no matter what the wind speed is.

Winds could be blowing 60 knots, but the paper will stick like glue to the table because the balloon and the table and the paper are all moving at 60 knots over the ground, but the balloon's relative speed in the air mass is zero. Or, to put it another way, the balloon's groundspeed is 60 knots and its airspeed is zero. When you move with the wind, you don't feel it.

Balloons float with the wind and have zero airspeed. Airplanes, on the other hand, need some airspeed to create a relative wind over the wings in order to create lift and fly. An airplane with zero airspeed can only go one direction, down.

Robinson Helicopter Company

Fig. 1-7. *Once an aircraft leaves the ground it becomes one with the wind: Robinson R22.*

An airplane cannot take off until it reaches a certain minimum airspeed. This airspeed varies according to aircraft type, weight, air temperature, and air pressure, but for the sake of an example, let's say that a particular airplane needs 60 knots airspeed before it can lift off from the ground.

If there were no wind at all, the airplane would have to accelerate along the runway until reaching 60 knots of airspeed and groundspeed because both are equal in a no-wind situation.

What happens if 40 knots of wind are blowing directly parallel to the runway?

With the airplane's nose pointed into the wind before the takeoff roll begins, its airspeed indicator would show 40 knots, even though its groundspeed is obviously zero. The airplane would only have to accelerate an additional 20 knots before it could take off; it would still take off with 60 knots airspeed, but its groundspeed would only be 20 knots.

And if the wind is 60 knots? The airplane would theoretically be able to fly directly over one spot, its airspeed 60 knots and its groundspeed zero—hovering.

Incidentally, all aircraft (except balloons) have airspeed indicators, airspeed being very important and relatively easy to measure. Groundspeed is actually irrelevant to an aircraft with respect to the physical act of flying. One's speed over the ground is of great relevance to navigation, however, and is much more difficult to determine, unless the craft is equipped with navigation equipment.

What's the point?

The point is that most airplanes, like the hypothetical one in our example, need nearly a hurricane-force relative wind before they can even take off. And after takeoff, they might accelerate to 100 or 200 or even 1,000 or more knots in the air. To sum up, airplanes are designed to fly in strong winds and require a strong relative wind to fly.

What about helicopters?

A helicopter doesn't need forward movement through the air to take off and land because the rotating wings create their own relative wind. Consequently, helicopters are able to take off and land with zero forward airspeed; however, after takeoff they can accelerate to airspeeds more than 100 knots.

Helicopters might not fly as fast as most airplanes—the fastest helicopter to date, a Westland Lynx, has flown at 216 knots but . . .

The fact is: Helicopters are designed and built to fly in strong winds.

This is not to say that strong winds do not create problems for aircraft. They do, but it is with respect to the ground that wind is a problem (FIG. 1-8).

The first problem with wind is during takeoff. An airplane is very sensitive to wind from the sides and back; therefore, even though a light airplane could take off

United Technologies Sikorsky Aircraft

Fig. 1-8. *Helicopters are built to fly in strong winds. It's with respect to the ground, such as during landing, that wind becomes a problem: Sikorsky SH-60 Seahawk.*

into a 60-knot wind, it might not be able to do so safely, particularly if the wind is not coming straight down the runway. Even if the wind is parallel to the runway, it might be impossible to taxi the airplane to the takeoff position without subjecting it to excessive crosswinds and tailwinds. It's not unusual for small planes to be flipped upside down because the pilot was not paying enough attention to the wind while taxiing around the airport.

Of course, a helicopter doesn't need a long runway and can be turned to face any direction in order to take off into the wind. However, starting up and shutting down a helicopter in high winds can be a problem. Until the rotors are turning at normal rotational speed, the blades are very susceptible to flapping up and down. If the wind is too high, a rotor blade can flap so low that it hits the top of the fuselage.

Another problem with high winds is fuel consumption. If an airplane or helicopter cruises at 120 knots and has a 60-knot tailwind, then there's no problem, groundspeed is 180 knots and it will get where it's going in 50 percent less time than if there were no wind. But if that 60-knot wind is right on the nose, the groundspeed will be only 60 knots and it's going to take it twice as long to cover the same distance than in the no-wind condition.

In both cases the airspeed is the same, 120 knots, so the rate of fuel consumption is the same. The difference is time. In the second case it might not be possible to load enough fuel in the tanks to make the trip. Theoretically, helicopters and airplanes could fly in any wind speed. They might not get anywhere and might be moving backward over the ground if the wind is greater than their maximum airspeed, but they could still fly.

The third problem with high winds is passenger safety when getting on and off the aircraft. This can be a serious consideration at offshore oil platforms where the wind is often greater than 40 knots.

Many North Sea helicopter operators have chosen 60 knots as the maximum wind speed limit at offshore oil platforms. This is for the safety of the passengers while boarding and disembarking from the helicopters. En route to the platforms, there is no wind limit as long as enough fuel for the round trip can be carried and the wind stays fewer than 60 knots over the platform helideck.

MYTH #4:
A flight in a helicopter is always
bumpier than a flight in an airplane

"Bumpier" usually means two things to helicopter passengers: vibrations and turbulence.

All right, I'll grant you that helicopters vibrate more than airplanes, particularly jet-powered airplanes. You'd vibrate, too, if you had has many moving parts as a helicopter. Even when a helicopter is as fine-tuned as a concert piano, vibrations occur. And, if just one thing is out of balance or adjustment, the ride in a helicopter can be very uncomfortable.

The first helicopter builders watched some of their inventions literally shake to pieces. Fortunately, the industry has come a long way since then. The subject of vibration reduction is one of prime concern among rotary-wing designers and manufacturers and, consequently, the newer helicopters vibrate a great deal less than their predecessors of only a few years ago.

Helicopter vibration might never be reduced to the level of a large passenger jet, but it's much better than it used to be.

With respect to turbulence, however, I take issue.

The fact is: Helicopters are more stable in turbulence than airplanes.

Just ask my wife, a reluctant airplane and helicopter passenger if ever there were one. She is one of the foremost authorities on turbulence. She hates it.

She always sits by the window in an airplane so she can watch the wing (to be sure it doesn't fall off) and to look out for clouds. She has learned from experience that clouds might make bumps. And she hates those bumps.

Before her first flight in a helicopter, my wife was very worried about turbulence. I tried to reassure her that the weather forecast was not that bad, but I had to be honest with her because it wasn't that good either.

We hit some turbulence during the flight and my wife was pleasantly surprised.

As I said before, the wings of a helicopter are its rotor blades. In flight, the blades can be considered a single unit, the rotor disc. Nearly all the weight of the helicopter hangs from the rotor disc like a giant dead weight. Although the fuselage itself creates some lift, it is the rotor disc that is doing the lion's share of the work.

Turbulence, which is defined as abrupt changes in the relative wind, causes disturbances to the flying part of an aircraft. Therefore, in a helicopter, it's the rotor disc that takes the brunt of any turbulence. By the time the effect of the turbulence is transmitted to the fuselage hanging below, it is dampened considerably by the various things that are built into helicopters to reduce vibrations. The sharp jolts and bumps often felt in an airplane become gentle to moderate humps in a helicopter.

My wife described turbulence in a helicopter as similar to the gentle rolling motion one experiences in a canoe when passing over the wake of another boat.

On the other hand, the wings of an airplane are mounted rigidly to the fuselage, as they must be. Any turbulence that affects the wings is directly transmitted to the fuselage and is felt by the passengers and crew.

To be honest, an average-size helicopter is not going to be as stable as a Boeing 747 in the same turbulence. The difference is the total weight; it's like comparing the ride in a canoe to that of an ocean liner. If you take an airplane and a helicopter of the same weight, you will have a more stable ride in the helicopter (FIG. 1-9).

The reason that helicopters seem bumpier than airplanes is relative and actual, relative because there are no helicopters as big as Boeing 747s. If you step off a 747 after crossing the Atlantic and then climb into an airport-to-city center helicopter, the helicopter is not only going to seem bumpier to you, it will be bumpier because it's a lot smaller.

Fig. 1-9. *The ride in a small helicopter is smoother than the ride in an airplane of equal gross weight: Schweizer 300.*

A flight in a helicopter over a certain route might actually be bumpier than a flight in an airplane over the same route because airplanes, even those as small as or smaller than helicopters, can often fly above the worst turbulence-producing storms and frontal systems. Helicopters, mainly because they lack icing protection, pressurization systems, and the power required to fly higher, have to muck along at lower altitudes and plow through the turbulence.

Things are changing. With the advent of such rotary-wing machines as the tilt-rotor, the improvement of conventional helicopter designs and engines, and the development of reliable helicopter anti-icing systems, cruising at altitudes above 10,000 feet will soon be available (FIG. 1-10). This will give the helicopter pilot much more flexibility in choosing an altitude that is less turbulent.

MYTH #5:
Helicopter pilots are different from other people

Thank you, Hollywood.

Not all helicopter pilots on television and in films are portrayed as being a little strange, but after seeing so many helicopters crash on the big screen and the tube, would any sane person want to be a helicopter pilot?

Bell Helicopter Textron

Fig. 1-10. *The tilt-rotor will be able to fly at higher altitudes where there is usually less turbulence than down below: Bell-Boeing V-22.*

Luckily for us, accidents are not nearly as prevalent in real life as they are in the movies and when they do occur, they're usually survivable.

And speaking of surviving, I think most pilots have a keener sense of self-preservation than most other people. This is perhaps because they respect the hazards inherent with flying and have to compensate for them on a daily basis (FIG. 1-11). Remember, it's always the pilots who are the first to arrive at the scene of an aircraft accident.

What does it take to be a pilot? An excerpt From OUTING Magazine is just as valid today as it was when it was written by Augustus Post in May, 1911:

> Flying . . . calls for the greatest exercise of self-control and requires, as essential elements for success, bravery, daring, to a slight degree, courage, confidence in yourself, your men, and your machine, good judgment, clear sight, intuitive knowledge, quickness of thought, positiveness of action, all combined with a most delicate sense of feeling and acute powers of perception. Good health is both a result and a prerequisite of good flying, and your mind must be clear and free.

Pilots start flying because they are lured by the thrill and adventure of the sky. They keep flying because they enjoy it, it's personally rewarding (FIG. 1-12), it pays well, or they get a fair amount of free time, or all four, if they're lucky.

Helicopter pilots consider themselves pilots first and helicopter pilots second. The reasons they fly helicopters instead of airplanes, or helicopters and airplanes, are as varied as the pilots themselves.

McDonnell Douglas Helicopter Company

Fig. 1-11. *Pilots know the hazards involved in flying and act accordingly. They don't go out of their way to take unnecessary risks: McDonnell Douglas AH-64 Apache.*

MBB Helicopter Corporation

Fig. 1-12. *Being able to help save lives is an important motivational factor for many emergency medical service (EMS) pilots: MBB BK117.*

I flew helicopters and airplanes while in the Air Force. I chose to fly helicopters because I got a bigger kick out of flying low over the ground than way up above the clouds and because the main mission of the Air Force helicopter fleet, at that time, was search and rescue. That appealed to me.

Other pilots began flying helicopters for different reasons.

The fact is: For all professional pilots, flying is a job.

And a job is something one likes to come home from.

When it comes right down to it, helicopter pilots really aren't that much different from other people. (I don't sound crazy, do I?)

They have wives and children, or husbands and children, or boyfriends or girlfriends, or girlfriends and children, or maybe a combination of the above, but not every combination. They might be single or married or divorced or separated, just like the rest of humanity. Some are deeply religious, some are less so, and others are atheists.

Helicopter pilots come home from their daily flying and play golf or wash dishes or watch television or have a couple of beers or take the dog for a walk or change a diaper or work on the house or run the family business or go fishing or work on their farm or help the kids with homework or cut the grass or raise horses or fiddle with a computer or chase girls or fix the car or write books.

The vast majority of pilots adhere fanatically to the regulations concerning consumption of alcohol and flying (in the United States, no booze eight hours before takeoff). Failure to do so can mean loss of license and job, but most follow the rule or extend the number of hours beyond eight because they are professionals and don't want to impair their own abilities when flying. Other pilots simply do not drink.

Very few, if any, professional pilots experiment with illegal drugs and I can't imagine a flight operator putting up with a pilot who did. Because the FAA has ordered mandatory drug testing for all professional pilots, it's doubtful any pilot who does take drugs would last very long before he was discovered.

Almost without exception pilots are rugged individualists and often politically conservative, but they usually have a strong sense of camaraderie and fellowship with other pilots. The Air Line Pilots Association is a strong union; ironically, attempts to unionize helicopter pilots have always failed in the United States. In many other countries, particularly in Europe, there are strong helicopter pilot organizations.

Most pilots take their physical condition seriously, although some smoke and many drink too much coffee. Eating unbalanced meals at odd hours while on the job is common; often pilots don't get enough sleep because of the nature of the job; many should probably exercise more. They still must pass a medical examination every six months or year, depending on their age and the regulations of the country. Without a valid medical certificate, a pilot is not allowed to fly.

Many people know that pilots need regular medical checkups, but many don't know that all pilots involved in scheduled passenger-carrying services, whether by helicopter or airplane, are required to pass a flight test and written examination every six months.

It's like taking a driving test twice a year, only believe me it's much more difficult. Most companies go a step further and require their pilots to have a few days training in conjunction with the flight check. Anything and everything having to do with that company's operations and specific aircraft operations is covered during this training.

The latest trend in training is the use of flight simulators (FIG. 1-13). There are companies whose sole purpose is to provide periodical simulator training to pilots. Such training is not required by law yet, but many flight operators send their pilots anyway, because they believe in the benefits of training in a simulator. Some companies send their pilots half-way around the world, just to get them into a simulator once or twice a year.

What can you do in a simulator? Just about anything done in the actual aircraft.

The inside looks exactly like the cockpit of a real aircraft and all the switches, instruments, warning lights, and controls work as if they were in the real aircraft. In the newer visual simulators, a computer-generated image fills the cockpit windows with simulations of airports, heliports, cities, countrysides, and offshore drilling platforms.

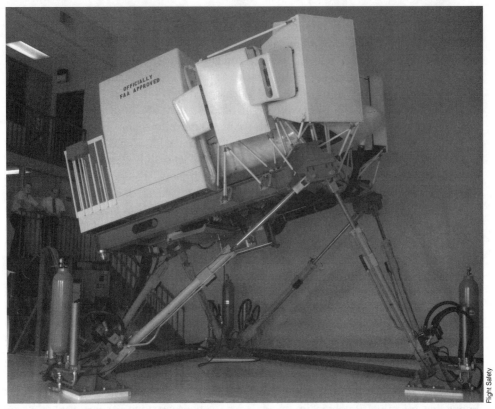

Fig. 1-13. *Flight simulators provide realistic initial and recurrent training for many helicopter pilots: Bell 222 simulator.*

Every conceivable emergency can be enacted in a simulator, including emergencies that are too dangerous or even impossible to simulate in the real aircraft. The ability to experience emergencies, to see the symptoms and warnings as they actually happen, to feel how the aircraft responds, and to learn and improve human reactions is a benefit of an aircraft simulator that every pilot respects. A pilot might not like the training as it is happening, but he appreciates its value afterwards.

Helicopter pilots might be a little different from other people, but they aren't the shell-shocked psychopaths often portrayed in movies. They are professionals who take pride in doing their jobs well.

So what do you think? Now that we've exposed the myths for what they are, do you still want to learn to fly helicopters? I hope so, because I think you'll find it a rewarding experience, whether you do it for a living or just for fun (FIG. 1-14).

The next chapter deals with one thing every helicopter pilot should know something about: aerodynamics. If the subject sounds technical to you, never fear. From

Aerospatiale AS 332 Super Puma

Fig. 1-14. *Learning to fly helicopters can lead to an interesting and rewarding career.*

personal experience, I've found that many ground and flight instructors try to give the student too much of what he doesn't need and too little of what he does; therefore, I've tried to simplify the more technical aspects about aerodynamics to reveal only what you need to know. Please don't skip the next chapter if you claim to be a nontechnical person. If I've done my job right, it shouldn't be difficult to understand.

Fair enough? Okay, let's go do it.

2
Basic Aerodynamics

Like all novices we began with the helicopter (in childhood) but soon saw it
had no future and dropped it. The helicopter does, with great labor, only
what the balloon does without labor, and is no more fitted than the balloon
for rapid horizontal flight. If its engine stops, it must fall with deathly
violence, for it can neither float like a balloon nor glide like an airplane. The
helicopter is much easier to design than an airplane, but it is worthless when
done.

Wilbur Wright
15 January 1909

CONVENTIONAL WISDOM STATES THAT ONE SHOULD KNOW SOMETHING
about the theory of flight before one actually begins to learn how to fly. This
makes sense to me, but some people might not agree.

Look at birds, for instance. They do it without going to ground school first.
Humans learn how to ride a bicycle without studying the principle of gyroscopic force
that maintains balance. Perhaps it is better to simply "kick the tires and light the
fires" and zoom off into the wild blue yonder with nary a thought about lift coeffi-
cients and thrust vectors and other such things.

On the other hand, many, if not most, of us older, more mature helicopter pilots
went through undergraduate pilot training with the United States Army, and the mili-

tary certainly does like to give briefings and classes to pilot candidates. As a result, I suppose I lean toward conventional wisdom in this case.

But theory doesn't have to be boring. So, don't skip this chapter just because you think you have a right-side-dominant brain.

LIFT AND AIRFOILS

Lift is a force that acts upon any object moving through air. Propel anything fast enough through the air and lift will be created. Depending upon the shape of the object you might also be creating an enormous amount of drag and expending great quantities of energy, but you will get some lift.

Lift is created by the air moving past the object (FIG. 2-1). As air moves over the object, the velocity of the air increases and the air pressure above the object decreases.

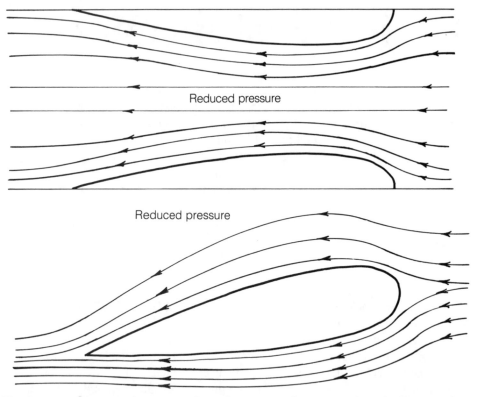

Reduced pressure

Reduced pressure

Fig. 2-1. *As a quantity of air moves through a restricted space, such as the Venturi tube in the upper drawing, the velocity of the air increases and its pressure decreases. Similarly, the velocity of the air moving over the top of the airfoil (in the lower drawing) increases and its pressure decreases. The higher air pressure under the airfoil causes the airfoil to move upward. This upward force is called lift.*

At the same time, the velocity of the air moving under the object decreases and the pressure increases; thus, the combination of decreased pressure on the upper surface and increased pressure on the lower surface results in an upward force. This is the force we call lift. For the scientifically minded, this principle is called the *venturi effect* or the *Bernoulli principle*.

Confused? Don't worry. You really don't have to understand why it works just as long as you believe it does work. In fact, the Bernoulli principle works whether you believe it or not. Perhaps the following experiment will help you understand how it works, if not why.

While riding in a car, stick your hand out the window. Hold your fingers together and your hand flat, palm toward the ground, thumb forward. You'll be able to find a neutral position easily, that position in which your hand will neither go up nor down in relationship to the ground.

Now rotate your hand slightly, turning the thumb-side upward. You'll feel the wind pushing your hand upward. If you could hold your hand at that angle and it wasn't attached to your arm, your hand would just keep on climbing, assuming, of course, that it continued to propel itself forward.

The more you increase the angle of your hand, the harder the wind pushes it up; in other words, the greater the force of lift acting on your hand (FIG. 2-2). If you try this experiment at different speeds, you'll discover that the faster you go the less you have to angle your hand in order to get the same amount of lifting force. In fact, all other things being equal, if the speed doubles, the amount of lift created increases four times. (Lift varies as the square of the velocity of an airfoil.)

Your hand is a simulated wing, or *airfoil*, what engineers like to call a wing. An airfoil is any surface designed to produce lift or thrust when air passes over it; therefore, airplane propeller blades and helicopter rotor blades are also airfoils.

In theory, almost anything can produce lift if it moves through the air fast enough, even anvils and bricks. But some things are obviously better lift-producers than others and anvils and bricks are definitely poor lift-producers. Actually, no matter how fast you propel anvils and bricks through the air you won't be able to make them fly because the lift they produce is not nearly enough to overcome their weight, but they do create some lift.

Airfoils, by definition, are good lift-producers and one of the most important tasks of an aircraft designer is to find the best airfoil for the aircraft in question (FIG. 2-3). Large, heavy transport aircraft need airfoils that are shaped differently from fast, high altitude fighters; helicopter airfoils are different from airplane airfoils.

A wing designed to fly faster than the speed of sound is not a good lift-producer at slower airspeeds and the opposite is true, too. An aircraft such as the F-111 was designed with movable swing wings so that the profile of the wing could be changed for slow or fast flight. This is very expensive and not practical for aircraft not funded by the military-industrial complex; therefore, aeronautical engineers always have to make trade-offs and compromises when designing airfoils.

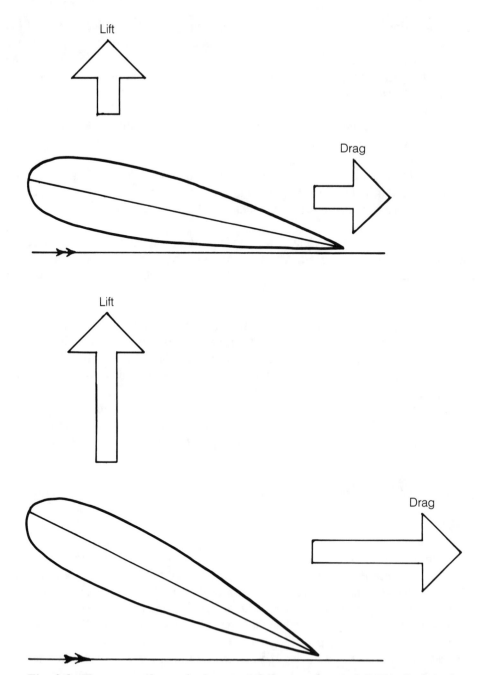

Fig. 2-2. *The greater the angle that an airfoil meets the wind (within limits), the greater the lift produced.*

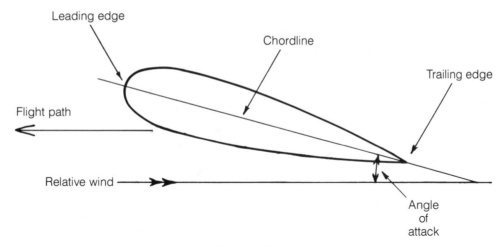

Fig. 2-3. *Elements of a basic airfoil. The airfoil in the drawing is symmetrical because the shape of the airfoil on both sides of the chordline is the same.*

The airfoils (rotor blades) on a helicopter create unique problems for the designer because helicopters not only fly forward, they must also hover and fly sideways and backwards. Rotor blades designed with good hover characteristics do not provide for great high-speed maneuvers, and vice versa. So, every helicopter rotor blade is a compromise that gives the machine good hover capability and good, but not too fast, forward flight capability.

STALLS

A limiting factor for every airfoil is stall. Most people relate the term stall to automobile engines and assume that when a pilot says his airplane stalled, he means the engine stopped. I told my wife I was going out to practice stalls in my airplane and she became very upset. "Do you have to really turn off the engine? Isn't that dangerous?" she asked.

I explained to her that I wasn't going to shut down the engine in flight. I was referring to the loss of streamlined airflow over the wings that causes a loss of lift and a large increase in drag (FIG. 2-4). When this happens, the airfoil literally stops flying and the airplane literally begins falling. Stalls are usually easy to get out of, if the pilot has enough altitude, but if he fails to manipulate the controls properly, he could cause the airplane to enter a spin. Spins are harder to get out of, require a good deal more altitude for recovery, and are rather frightening the first few times you practice them.

An airplane can stall at high airspeeds and even when the engines are providing full power. This is rare. Most unintentional stalls in airplanes happen at low airspeeds and low power, usually during takeoff or landing. Much of a fixed-wing pilot's train-

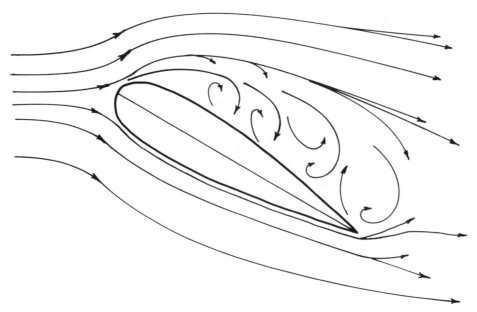

Fig. 2-4. *Air flow over an airfoil in a stalled condition. Airstream break up on the upper side of the wing causes a large decrease in lift and increase in drag.*

ing deals with how to recognize the approach of a stall, how to avoid it, and what to do after the aircraft has entered it. The first thing any experienced pilot will do when checking out in a new airplane is fly up to a safe altitude and do a series of stalls in various configurations (flaps up, flaps down, with power, without power) just to learn how the airplane reacts under these conditions.

Most airplanes have some kind of stall warning device to help pilots recognize the approach of a stall. This can be as simple as a horn or a voice warning that starts to sound as the airplane begins to stall. In large airplanes, a mechanical device called a *stick-shaker*, shakes the control column, activates before the airplane actually enters a stall.

Helicopter airfoils can stall, but not in the same way as airplanes, although the same aerodynamic principles still apply. Because a helicopter's wings (rotor blades) are always moving quite fast, a helicopter cannot enter a low airspeed stall as an airplane does, no matter how slow the helicopter's forward airspeed is, even zero. If a helicopter stalled at slow airspeeds such as an airplane, it couldn't hover. This ability to fly at slow airspeeds and hover without stalling is one of the main advantages helicopters have over airplanes.

Retreating blade stall

On the other hand, a helicopter will enter a high-speed stall at forward airspeeds that are much lower than are common for airplanes. Again, this is due to the helicopter's rotating wings.

Imagine a hovering helicopter in a no-wind condition, the blades turning counter-clockwise (FIG. 2-5). For the sake of simplicity, also imagine four main rotor blades at this precise moment located at the 90-, 180-, 270-, and 360-degree positions with respect to the fuselage. It's easy to understand that the *tip speed* of each rotor blade (the airspeed at the outer tip of the blade) is equal.

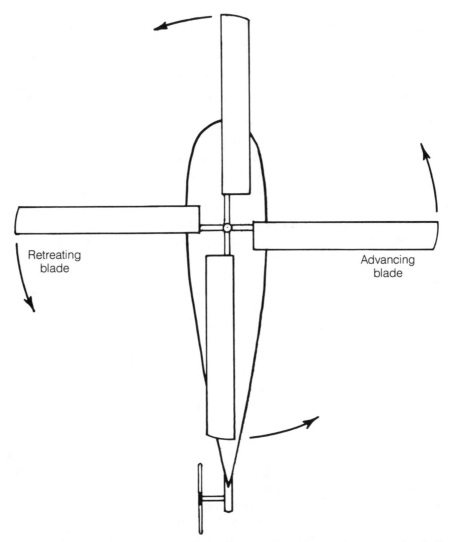

Fig. 2-5. *If a helicopter is hovering in a no-wind condition, the tip speeds of all the blades are equal, for example, 400 knots. If the helicopter is moving forward at 100 knots, the tip speeds are unequal. The advancing blade has a tip speed of 500 knots and the retreating blade has a tip speed of only 300 knots. The blades at the 360- and 180-degree positions have tip speeds of 400 knots.*

Now consider the same helicopter in forward flight.

The blade on the right-hand side, at 90 degrees, is meeting the airflow head on, as if your hand were sticking out of a car window. That blade's tip speed is the sum of the blade's rotational speed plus the forward airspeed of the entire helicopter. This blade is called the *advancing* blade because it is advancing into the wind.

The blades directly at the 360-degree and 180-degree positions have airspeeds equal to only the rotational speed, because the forward speed of the helicopter is not contributing anything to these blades' total airspeed.

The total airspeed of the rotor blade in the 270-degree position is less than its rotational speed because it is moving in the opposite direction of the helicopter. This is the *retreating* blade because it is retreating from the forward airflow; therefore, its airspeed is equal to the rotational speed minus the forward speed of the helicopter and as a result has the lowest total airspeed of all the blades.

Remember, the rotational speed of the rotor blades is constant, or nearly so, and that each individual blade sequentially becomes the advancing and the retreating blade as it rotates. It is the position of each blade in relationship to the direction of movement that causes the changes in their total airspeed.

Perhaps you're getting some idea of the complexity of the problems that faced and still face helicopter designers.

Think about the lift experiment in the car again. As the car's speed increased (and the airspeed acting on your hand increased), you needed less and less angle to create the same amount of lift. You can see the general principle involved here: To create an equal amount of lift, you need higher angles at low airspeeds than you need at high airspeeds. Likewise, if two airfoils are meeting the air at the same angle, the airfoil with the greater speed will be producing more lift. Unfortunately, the analogy of a hand outside a car, an airplane, and a helicopter breaks down because of the different ways airplanes and helicopters create lift and thrust.

Recall basic aerodynamic theory that the four major forces acting on any aircraft are *lift, weight, thrust,* and *drag* (FIG. 2-6). Lift is the force that acts upward and, as we have discussed, is the result of air flowing over an airfoil. Weight is the force that acts downward and is as simple as it sounds, the weight of the aircraft. Thrust is the forward-acting force that propels the aircraft forward. Drag opposes thrust and is a combination of air resistance and inertia.

To fly faster in an airplane while maintaining a level altitude, a pilot must increase thrust by increasing engine power. In a propeller-driven airplane, the propeller blades rotate faster; in a jet-powered airplane, the turbine blades inside the engine rotate faster. The wings' angle of attack remains constant.

A helicopter creates lift and thrust with the main rotor system. When hovering, the helicopter as a whole has no thrust and no drag; all thrust created by the main rotors is acting vertically as lift to oppose weight; however, each blade is creating lift (and drag) by virtue of its movement through the air. To fly forward in a helicopter, part of this vertical thrust/lift vector must be directed forward in order to produce a horizontal thrust vector.

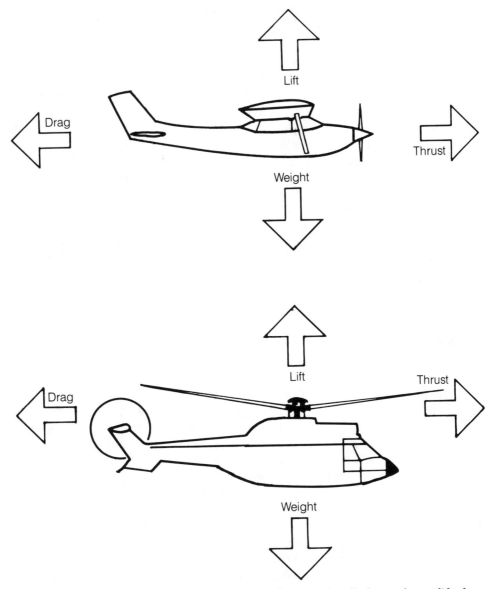

Fig. 2-6. *The four major forces acting on a powered aircraft in flight are thrust, lift, drag, and weight. The main rotor blades of a helicopter provide thrust and lift.*

Engine power must be increased to fly faster in a helicopter, as in an airplane; however, unlike an airplane's propeller, a helicopter's rotor blades do not rotate faster to create more thrust. Instead, the pitch angle on all the blades is increased in order to increase overall thrust/lift. By tilting the entire rotor disc progressively more forward, more and more of this total thrust/lift vector is converted to forward thrust.

The joker in all of this is the rotational speed of the rotor blades, called *rotor speed* or *rotor rpm*. Generally, every helicopter has an optimal rotor speed and therefore this speed should remain as constant as possible. But when you increase the pitch of the rotor blades, you not only increase lift, but drag, too. This means you must use more engine power in order to keep the rotor blades rotating at the same speed.

Let's try another analogy: swimming. Imagine doing the breast stroke with your hands flat in the water (parallel to the surface) at one stroke per second. It doesn't take much effort to swim this way, but you don't move very fast either, if at all. Now, angle your hands slightly and pull at the same rate. This takes more energy, but your body now moves through the water. Angle your hands 90 degrees to the surface of the water and it requires a lot of energy to maintain the same rate of one stroke per second, but your body quickly slides forward.

This is similar to what a helicopter has to do. The angle of the main rotor blades must increase to move the helicopter faster through the air while maintaining the same rotational speed. But there's an upper limit to this because the blade moving the same direction as the helicopter (the advancing blade) is moving faster than the blade moving opposite to the forward direction (the retreating blade) and therefore creating more lift. If this *dissymmetry of lift* is not compensated for, it would be impossible for a helicopter to do anything but hover.

The problem was first solved by Juan de la Cierva with his autogiros and his solution is considered his greatest contribution to the eventual development of true helicopters (FIG. 2-7). De la Cierva mounted his autogiro blades on hinges that allowed them to flap individually within set limits (FIG. 2-8).

Amazingly, this flapping action automatically compensated for the dissymmetry of lift by allowing the advancing blade to flap up and the retreating blade to flap down. The upward flapping of the advancing blade causes its effective area of lift and angle of attack to be reduced while the downward flapping of the retreating blade causes its effective area of lift and angle of attack to be increased. The overall effect is to equalize lift across the entire rotor disk. It might sound too good to be true and a bit hard to comprehend, but it does work.

The down side of this flapping action is the fact that the retreating blade always must have a higher angle of attack than the advancing blade. As said before, a limiting factor for every airfoil is stall; in other words, every airfoil has a stall angle of attack, which, when exceeded, causes the blade to stall.

Because the retreating blade's angle of attack is always higher than the advancing blade's angle of attack in forward flight, the retreating blade is always the first to stall. (Remember that each blade successively becomes the retreating, then the advancing, then the retreating blade, over and over again as it rotates.) As the air stalls on the retreating-blade side of the helicopter, an imbalance is created and high lift on the advancing side and low lift on the retreating side causes the helicopter to pitch up and roll toward the retreating side. This is known as *retreating blade stall*.

Fortunately, retreating blade stall is easy to avoid, easy to get out of, and therefore not nearly as dangerous or encountered as often, as stalls in airplanes. Helicopter

Fig. 2-7. *Juan de la Cierva's greatest contribution to rotary-wing flight was his development of the flapping hinge for his autogiros: Pitcairn Autogyro, 1930.*

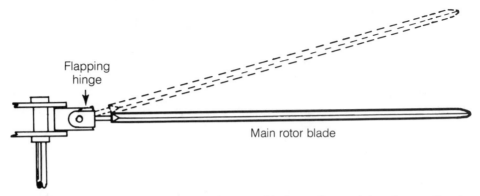

Fig. 2-8. *The flapping hinge permits the rotor blades to flap and thereby equalize the lift between the advancing half and the retreating half of the rotor disc.*

manufacturers must establish maximum airspeed limits that are below the threshold of retreating blade stall for each helicopter they produce. Pilots who inadvertently encounter retreating blade stall know they must reduce the collective pitch (*see* index for explanatory reference to collective pitch) of the main rotor blades to reduce the angle of attack on all the blades and thereby bring the retreating blade out of the stall region.

Retreating blade stall is so rare that civilian pilots just about never encounter it because they rarely fly close to the maximum allowable airspeed of their helicopters. Military pilots who must, out of necessity, push their machines to the limits of the operating envelope, perhaps occasionally feel the onset of retreating blade stall (FIG. 2-9). They always keep it in mind but probably don't lose any sleep worrying about it.

United Technologies Sikorsky Aircraft

Fig. 2-9. *Retreating blade stall is more of a concern to military pilots who sometimes must push their aircraft to the limits of the operating envelope: Sikorsky MH-60 Pave Hawk.*

Settling with power

Another more common way a helicopter can "stall," *settling with power*, occurs during landings with very light wind or tailwind conditions when the descent angle is vertical or nearly so.

Anyone who has stood under or near a hovering helicopter knows that it creates quite a wind. Not all the *downwash* goes straight down. The air disturbed by the outer rim of the rotor disc goes down, too, and then curls up in a circular pattern. It is similar to and very much akin to the vortex-shaped wake turbulence that is created by airplanes.

If a pilot allows the helicopter to descend too fast vertically, the main rotor begins to encounter its own turbulent downwash (FIG. 2-10). The turbulent air stalls over all the rotor blades, reducing lift, and the helicopter descends faster. If the pilot tries to correct the increased rate of descent by increasing power, he aggravates the situation by making the downwash more turbulent and the rate of descent only increases.

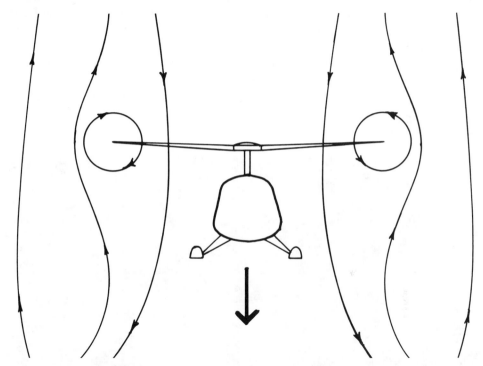

Fig. 2-10. *Airflow pattern when a helicopter is settling with power. The condition is characterized by a high rate of descent that increases even more when more power is applied to the helicopter (when collective is increased).*

Settling with power can be hazardous if it happens too close to the ground and the pilot does not recognize it soon enough. Fortunately, getting out of settling with power is not too difficult if the helicopter is not too low. The pilot only needs to maneuver his craft out of the column of turbulent air; he pushes the nose down to pick up forward speed and reduces the collective pitch on the blades.

Settling with power in a helicopter is similar to a low-speed stall in an airplane. It is something that helicopter pilots are trained to avoid.

TORQUE AND TAIL ROTORS

A chapter about helicopter aerodynamics wouldn't be complete without a few words about torque and the various ways helicopter designers have tried to counteract its effect.

The reason conventional helicopters have tail rotors is to counteract torque effect. A portion of Newton's law of motion states: "For every action there is an equal and opposite reaction." Without a tail rotor, the main rotor blades would turn one direction and the fuselage would turn the other direction.

It's perhaps unfair to classify helicopters with one main and one tail rotor as "conventional" (FIG. 2-11) and everything else as nonstandard or unconventional. After all, there are a lot of helicopters without tail rotors that fly as well as or better than those with them. And the tail rotor is certainly no holy cow. It's sort of an add-on, a necessary evil put there to counteract the torque of the main rotor. Not surprisingly, then, most, if not all, of the so-called unconventional designs were motivated by an attempt to get rid of or modify the pesky tail rotor.

A single main-rotor helicopter must have a tail rotor, without it the helicopter is unflyable. But the tail rotor doesn't do its work for free. In fact, it extracts the proverbial "pound of flesh" in more ways than one.

Bell Helicopter Textron

Fig. 2-11. *The Bell 47, the first helicopter to be awarded a commercial license in the United States, is a conventional helicopter. The three-seat version shown here came afterward.*

First, it adds weight and a long tail that has no other purpose than to support the tail rotor. Much of the tailboom is hollow and cannot be used for equipment or baggage because it would cause problems with the fore and aft center of gravity.

Second, the tail rotor requires power that could otherwise go to the main rotor and therefore reduces the lifting capability of the helicopter. The tail rotor wastes lift horizontally.

Third, tail rotors are dangerous to living things. Tail rotors spin so fast they become virtually invisible; in certain light conditions, they are invisible. Loading a helicopter from the rear while the engine is operating must be done very carefully. The ducted-fan tail rotor reduces this particular disadvantage of tail rotors, but not the other problems (FIG. 2-12).

Fig. 2-12. *The Boeing Sikorsky light helicopter demonstrator aircraft is a modified Sikorsky H-76 with a ducted-fan instead of a conventional tail rotor.*

Fourth, the tail rotor and its support boom are limiting factors when the pilot is maneuvering the helicopter in very confined areas.

Finally, tail rotors limit the operating capability of the helicopter. It is possible in strong crosswinds for the pilot to run out of tail rotor, or yaw, control authority. This happens when the antitorque required from the tail rotor exceeds the amount of force it can provide. With full pedal deflection, a pilot simply cannot keep the nose pointed straight. For this reason, all conventional helicopters have crosswind limits considerably lower than their forward speed limit.

UNCONVENTIONAL HELICOPTER DESIGNS

Helicopter designers have known about the tail rotor's disadvantages for a long time and they have tried many ways to eliminate it. The most obvious design was to have two main rotors that turn in opposite directions and thereby cancel out the torque effect of each other.

Twin-rotor helicopters use all available engine power for lifting. None of the power is wasted to counteract rotor torque; furthermore, twin-rotor helicopters are not as "yaw control limited" as single-rotor machines; they can tolerate much higher crosswinds. Finally, with both rotors horizontal and higher off the ground, twin-rotor helicopters aren't as dangerous for people on the ground.

A twin-rotor helicopter can be built three ways:

- Side-by-side rotors.
- Tandem rotors.
- Coaxial rotors.

The side-by-side (lateral) design has been tried often, frequently with success. The FW 61 piloted by Hanna Reitsch inside the Deutschlandhalle in Berlin in 1938 had two lateral rotors. The largest helicopter ever built, the Mi-12 with a loaded weight of more than 230,000 pounds, also had lateral rotors. The design obviously works but it has a few disadvantages. For example, placing the main rotors on the sides increases the width and adds complexity and weight (an additional gearbox and strengthening selected parts).

Modern technology often stimulates the revival of older designs. Considering the fact that the tilt-rotor can be a side-by-side rotor helicopter, which it is when in helicopter configuration, then we'll probably be seeing a renaissance of lateral rotor helicopters before the end of the century.

The tilt-rotor is a hybrid; it can fly as fast as a turboprop airplane and hover like a helicopter, which is possible with tiltable main rotors on the wings. When hovering, the rotors are horizontal; in forward flight, they can be tilted 90 degrees forward for airplane conversion.

Although the idea is not new, it was not until the 1980s that the tilt-rotor became an operational possibility. The Bell V-22 (FIG. 1-10) is being tested for the United States military and a commercial version is in the works.

It's also possible to build a side-by-side twin rotor helicopter with the rotors so close that they intermesh. The Kamam H-43 Huskie with a side-by-side was flown for many years by the United States Air Force for close-to-airport rescue service.

Seeing the rotors intermesh was disconcerting, but in flight the Huskie was no different from other helicopters. One big advantage was the large clam-shell doors in the rear that could be used with no fear of encountering a tail rotor. It was used successfully for many years until obsolescence: too small, too slow, and too short a range for Air Force needs.

The side-by-side design works well, but the tandem rotor helicopters have been more popular. Frank Piasecki built and flew a successful model in 1945. Numerous other manufacturers have produced tandem-rotor helicopters: Bristol, Vertol, and Sikorsky among others. Boeing's 234 is the only tandem-rotor helicopter in production, but it has had limited success due to the declining offshore oil market and accidents. Its military forebearer, the CH-47, remains quite popular with armed forces all over the world and has been in continuous production for over 20 years (FIG. 2-13).

Fig. 2-13. *The Boeing MH-47E is a tandem rotor helicopter.*

Mounting one rotor above the other is another way to accommodate two main rotors rotating in opposite directions. This requires two main rotor shafts, one rotating inside the other one; hence, the term *coaxial contrarotating rotors*. The design is rather complex and heavy, but workable. The Kamov design center in the Soviet Union has built many of these helicopters for military and civilian roles (FIG. 2-14).

Fig. 2-14. *The Kamov KA-32 is an example of a helicopter with coaxial contrarotating rotors. Notice the complex design of the main rotor head.*

A more exotic way to eliminate the tail rotor is to use a *jet tip* helicopter. Propulsion on the tips of the rotor blades avoids the torque problem altogether. The idea is intriguing and has attracted countless researchers. Many types of jet propulsion schemes have been tried, including rockets, ramjets, turbojets, and compressed air. Some have even flown; however, all tip jet helicopters have three things in common: high noise level, poor fuel-efficiency, and concept abandonment. Who knows? Perhaps technological advances in the future will make jet tip helicopters more attractive.

The X-wing helicopter derivative, another exotic design, had to wait for advances in four key technologies:

- Circulation control airfoils.
- Fully redundant digital fly-by-wire control systems.
- Pneumodynamic valving systems.
- Composite structural materials.

The experimental aircraft took advantage of the *Coanda effect* by using a circulation-controlled airfoil to create lift by blowing pressurized air out of one or more slots in the airfoil. The Coanda effect was named for Henri Coanda who discovered the phenomenon in 1910; the effect is the tendency of an airflow to stick to a curved surface if pulled along by a dynamic force.

The X-wing objective was to create an aircraft that could take off and land vertically and fly more than 400 knots. Sikorsky Aircraft and NASA worked jointly on the X-Wing until funds were depleted in 1988.

Interestingly, one of the most commercially promising possibilities of eliminating the tail rotor also uses the Coanda effect. McDonnell Douglas has developed the NOTAR helicopter and in 1989 announced plans to build and sell two models (FIG. 2-15).

Instead of a tail rotor to counteract the torque of the main rotor, NOTAR uses a turbine fan to pressurize the air inside the tailboom. Slots in the tailboom release a portion of the air to cause the main rotor downwash to adhere to the tailboom and create a sideward lift vector. The curved surface of the tailboom—the NOTAR tailboom has a considerably wider diameter than a conventional tailboom—is an airfoil and uses the Coanda effect to enhance the antitorque force. Additional force comes from venting the remaining air through a rotatable direct-jet thruster at the end of the tailboom. Antitorque pedals in the NOTAR work the same as tail rotor pedals in a helicopter with a conventional tail rotor.

ALL ELSE, AERODYNAMICALLY

Many other aerodynamic principles affect helicopters: gyroscopic precession, centrifugal force, blade coning, Coriolis force, among others. A pilot is primarily concerned about the principles discussed so far, plus a few more that shall be explained in subsequent chapters.

Fig. 2-15. *An MD 520N, equipped with a NOTAR antitorque system, hovers in the glow of an Arizona sunrise.*

Additional aerodynamic principles that affect helicopters are defined in the glossary. In 20 years of helicopter flying, the only time I really had to worry about the more esoteric aerodynamic stuff was when I took written tests for the Army, Air Force, and FAA. Refer to supplemental material if you want to know more about helicopter aerodynamics, beyond the basics.

Aerodynamic theory is great for the armchair pilot, but when it comes to really flying, what you do with your hands and feet determines how you translate theory into action. The next chapter explains helicopter flight controls systems, the tools that connect the pilot to Bernoulli's principle, and that other aerodynamic stuff.

3
Flight Controls

To study the action and permit ourselves some training, we mounted the helicopter without the main rotor on a support that permitted tilting the machine in all directions. The controls appeared satisfactory and, after about one week of training, the main lifting screw was mounted again and flights were resumed. In flight it was much better, but we were more careful now because we had learned that with inadequate control and experience the aircraft could easily turn over near the ground—as it did once, before the new control system was installed.

Igor Sikorsky and the VS 300 in 1940, from his book "The Winged S"

HELICOPTERS HAVE FOUR FLIGHT CONTROLS: COLLECTIVE, THROTTLE, cyclic, and tail rotor pedals (FIGS. 3-1 AND 3-2). When you fly, you'll usually be manipulating two or more of these in combination during most maneuvers. Chapter 11 delves into aircraft systems in more detail, but for now let's look at each of the flight controls separately and find out what it does.

THE COLLECTIVE

The collective control is the stick or lever that the pilot moves with his left hand. Basically, the collective changes the pitch, or angle, on all the blades exactly the same

Fig. 3-1. *Cockpit of Schweizer 300 showing the location of the cyclic, the collective, the throttle control, and the tail rotor pedals.*

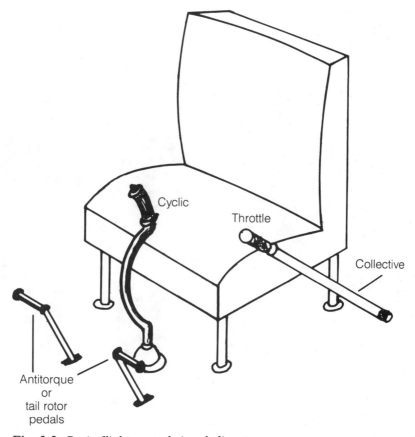

Fig. 3-2. *Basic flight controls in a helicopter.*

amount. Or, more precisely, the collective pitch lever changes the pitch on all the blades collectively.

Recall the swimming example in chapter 2.

When you do the breast stroke, you instinctively equalize the angle that your hands meet the water; you don't hold one hand flat and the other one perpendicular to the surface; if you did, you'd go around in circles. A helicopter needs a mechanism to ensure that an equal amount of pitch can be demanded from each of the blades. This is done with the collective pitch lever and a series of rods, bellcranks, swashplates, and other things I won't get into yet.

The easiest example to visualize is a helicopter hovering in a no-wind condition (FIG. 3-3). If the pilot wants to climb straight up, he needs to increase the pitch on all the blades an equal amount simultaneously. He pulls the collective lever up slightly, increasing the pitch (or angle) on all the blades, which causes the angle of attack of each blade to increase, which means more lift is created by each blade equally

MBB Helicopter Corporation

Fig. 3-3. *The collective changes the pitch of all the rotor blades an equal amount. An increase in collective will cause a helicopter in a hover to ascend vertically; a decrease in collective will cause a helicopter in a hover to descend vertically: Bavarian Police Helicopter Squadron BO 105CB.*

(because all the blades are rotating at the same speed), and as a result total lift increases, too. The helicopter moves upward.

To descend, the pilot pushes the collective down, and the opposite happens. The pitch angle on all the blades decreases, which causes the angle of attack of each blade to decrease, which means less lift is created by each blade and therefore total lift decreases, too. The helicopter goes down.

This is the theory, anyway, but there is one important thing that must be compensated for or the helicopter won't work at all. This is aerodynamic drag.

As you increase the angle of attack of an airfoil to increase lift, you also increase drag (FIG. 2-2). In a helicopter, this means as you increase the pitch angle on the blades, you make it harder for the blades to move through the air and the blades' rotational velocity, the *rotor revolutions per minute* (rotor rpm), will decrease.

It's similar to trying to cut through butter with a knife. If you use the edge of the blade, the knife will slice right through, but if you use the flat side of the blade, it takes a lot more pressure.

Of course, the opposite happens if you reduce the collective pitch: It's now easier for the blades to move through the air and the rotational speed will increase. Because the rotor system of a helicopter is designed to operate at a certain optimal rotational speed—bad things start to happen if the rotor rpm is too low or too high—it is important to keep the rotational speed constant, or at least within rather narrow limits.

This has to be done with engine power. In the simpler helicopters with piston engines, and in some helicopters with turbine engines, a throttle grip is installed on the collective lever. This throttle grip is not unlike the throttle grip on a motorcycle.

A helicopter's throttle is controlled with the left hand—a motorcycle throttle is controlled with the right hand—and to increase power a helicopter throttle must be rotated in the opposite direction of a motorcycle throttle. This is one of the reasons people who have ridden motorcycles a great deal sometimes have a problem learning to fly helicopters.

THE THROTTLE

One of the first things a new helicopter pilot learns is that when he pulls the collective up he must also add power to keep rotor rpm from decreasing and as he lowers the collective he must reduce power to keep rotor rpm from increasing. In fact, my first homework assignment from my Army flight instructor was to do 500 "power on/power off" actions with my left hand to try and imprint the technique into my brain early in training (FIG. 3-4).

Imagine you have a short rod in your left hand to practice a "power on/power off." Or better yet, get a broomstick. Sit in a chair and lay the broom on the floor on your left side, with the top of the broom near your left foot and the bristles behind you. Lean over and grasp the top of the broomstick. You now have a simulated collective.

Sit up straight and hold the broom at arm's length. Now, lift the broom upwards, bending your elbow and rotating your knuckles toward your body. (Keep the bristle end of the broom on the floor.) This is the action of a "power on." Say, "Power on," out loud and turn your wrist as far as you can. Next, rotate your knuckles away from your body while you lower the broomstick, straighten your arm, and say, "Power off." You'll find the movement of your arm and wrist completely natural, very similar to doing an inside curl with a dumbbell. If you had to twist your wrist the opposite direction as you raise your arm, the action would be quite unnatural.

Newer, sophisticated turbine-engine helicopters have eliminated the throttle from the collective pitch lever and replaced the throttle with electronic devices that automatically change the power required to maintain rotor rpm whenever the pilot changes the position of the collective (FIG. 3-5).

Many turbine-powered helicopters have the engine speed levers on the center overhead panel. Once you move them to a "flight" position after starting the engines, you don't have to touch them until you want to shut the engines down. This has reduced the helicopter pilot's workload considerably.

Unless you happen to be independently wealthy or Uncle Sugar is paying for your helicopter training, you'll probably start out on one of the small, piston-powered helicopters that still has a throttle. So, practice those "power on/power offs."

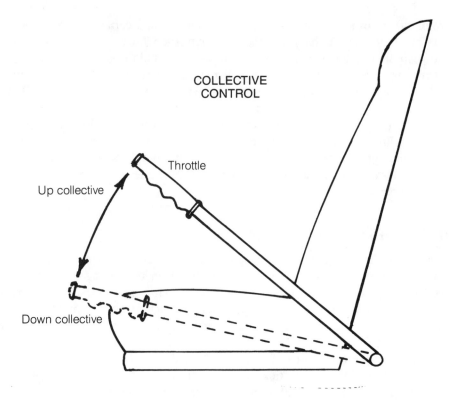

COLLECTIVE
CONTROL

Throttle

Up collective

Down collective

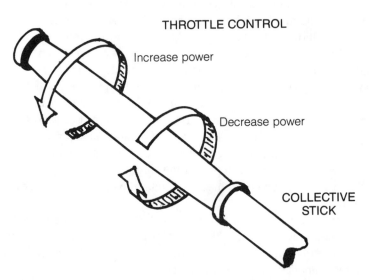

THROTTLE CONTROL

Increase power

Decrease power

COLLECTIVE
STICK

Fig. 3-4. *Changes in collective pitch must be coordinated with adjustments to the throttle. Practice "power on/power offs" until the action becomes second nature.*

Fig. 3-5. *The Boeing 234 collective does not have a motorcycle-type throttle. The two ridged switches on the left side are used to make fine adjustments to engine rpm.*

Even though small, piston-powered helicopters don't have the sophisticated electronic devices that adjust engine power to maintain rotor rpm, they do have mechanical devices linking the throttle to the collective. This feature, called *correlation*, automatically increases power when the collective is raised and decreases power when it is lowered. It's a good feature, but the pilot still must usually make fine adjustments with the throttle to keep rotor rpm precisely where he wants it.

THE CYCLIC STICK

A helicopter pilot's right hand holds and adjusts the cyclic stick (FIG. 3-6). When hovering, the cyclic is used to move the helicopter forward, backward, and sideways. When flying forward, the cyclic is used to bank into turns and, together with the collective, to climb, to descend, and to adjust the airspeed.

Mechanically, the cyclic changes the pitch on the main rotor blades in only one portion of the entire rotor disc. Consequently, lift is increased on one side of the disc and decreased on the other. As each blade rotates around the rotor mast, its pitch increases and decreases as determined by the cyclic stick position. The control is called the cyclic stick because the pitch on each blade changes as it cycles around the rotor disc.

Fig. 3-6. *Flight control placement in four helicopters: (A) Bell 206B JetRanger,*

(B) Enstrom F-28F Falcon,

(C) Aerospatiale AS 332 Super Puma,

Agusta Aerospace Corporation

(D) Agusta A109C.

For example, if you want to move toward the right while hovering, you must decrease the lift on the right side of the rotor disc and increase lift on the left side so that the resulting lift vector pulls the helicopter toward the right. To be 100 percent correct, there's actually a 90-degree lag in the lift vector due to gyroscopic precession of the rotor system; therefore, the changes in lift must occur not left and right, but forward and aft to move sideways.

Fortunately, this phenomenon, which is a characteristic of all rotating bodies, was discovered early in the game by pioneer helicopter designers and engineers. They very wisely decided to rig the controls such that when the pilot moves the stick to the right, the helicopter moves right and when he moves it left, the helicopter moves left.

As a result, the pilot doesn't have to think about gyroscopic precession or where the lift vectors are because these have been taken care of by the designers and engineers. And thank God for that because it sure would be hard to fly a helicopter if you had to move the stick forward to make it go to the right . . . or would you have to move it right to make it go forward? Don't worry about it.

Actually, the cyclic is very user-friendly: move the cyclic forward and the helicopter moves forward, move it backward and the helicopter moves backward, move it right and the helicopter moves right, move it left and the helicopter moves left.

What could be simpler?

THE TAIL ROTOR PEDALS

The fourth control the helicopter pilot must contend with is the tail rotor pedals. Also called antitorque pedals, these resemble the rudder pedals in an airplane and have the same function, yaw control, but work in a different manner.

A single main-rotor helicopter, which is often called a conventional helicopter, must have a tail rotor in order to counteract the torque created by the main rotor. Without a tail rotor, the main rotor blades would spin one way and the fuselage would spin the other way and flying in a helicopter would be impossible.

The tail rotor creates a force, lift, in a direction opposite to the torque effect of the main rotor (FIG. 3-7). Essentially it is a small main rotor mounted vertically instead of horizontally, pushing the tail one way so that the nose of the helicopter doesn't go the other way. A tail rotor is mounted far out on the boom to increase effectiveness with a longer moment arm.

(I suppose one could mount the tail rotor in the front of the main rotor — after all, helicopters can fly backwards with the tail rotor leading the way; however, they don't fly backward as quickly or as smoothly as they fly forward and sticking the thing in front of the cockpit would obviously obstruct a pilot's forward view. It would also have rather severe aerodynamic consequences, look funny, and have to be called a nose rotor.)

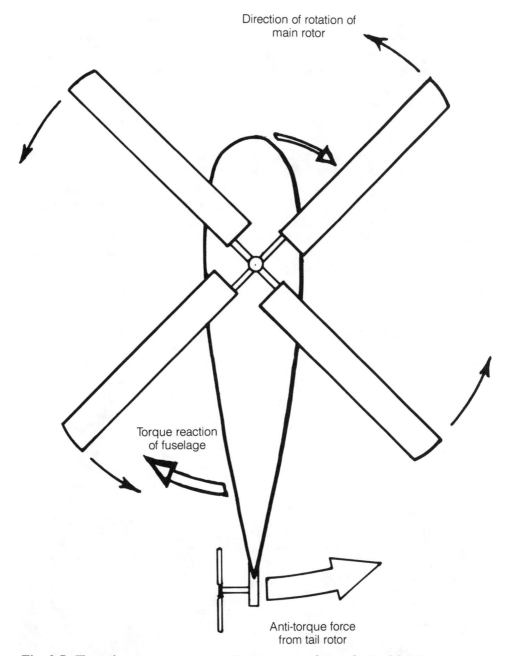

Direction of rotation of
main rotor

Torque reaction
of fuselage

Anti-torque force
from tail rotor

Fig. 3-7. *The tail rotor counteracts main rotor torque by producing lift sideways.*

The torque effect of the main rotor is not constant. It changes whenever there's a change in collective pitch, power, airspeed, and cyclic. In other words, torque effect is always changing, except in the most stable conditions; therefore, it's necessary to provide some way to control the antitorque force of the tail rotor.

One way to do this would be to change the rotational speed of the tail rotor, in other words make it spin faster or slower, but this would be very impractical. The tail rotor is connected directly to the main gearbox by a long shaft so that its rotational speed is always in direct proportion to the rotational speed of the main rotor. (Tail rotors normally spin four to five times faster than the main rotor.) Increasing and decreasing tail rotor speed is not easy to do.

The other way to vary the power of the tail rotor is to change the pitch angle of the blades. Because there is no need for cyclical changes of the tail rotor, the blade angles change by equal amounts, collectively. Pushing on one tail rotor pedal (left or right, depending on which side of the fuselage the tail rotor is mounted and the direction of rotation of the main rotor) causes the pitch angle on each tail rotor blade to increase; therefore, the lift generated by the tail rotor also increases. Pushing on the other tail rotor pedal causes the pitch angle to decrease; therefore, the antitorque force decreases.

Fig. 3-8. *The tail rotor pedals change the pitch of the tail rotor blades collectively so that the lift produced by the tail rotor can be varied: Robinson R-22 Beta tail rotor.*

In essence, the tail rotor pedals change the pitch of the tail rotor blades in the same manner that the collective changes the pitch of the main rotor blades (FIG. 3-8).

A helicopter pilot knows all this, of course, but he doesn't have to think about it much. What he does need to remember is that to make the nose of the helicopter turn to the right, he must push on the right pedal and to make it turn left, he must push on the left pedal.

This sounds very simple and it is; however, when first learning to hover a helicopter it takes a tremendous amount of concentration. Any small change in any one of the controls—collective, throttle, cyclic, and pedals—causes an adjustment in at least one other control, and sometimes all three other controls. A small change in wind can do the same thing.

ALL TOGETHER NOW

Suppose you are hovering a helicopter and want to move to the right a few feet (FIG. 3-9). To move right, you remember to move the cyclic stick slightly to the right. Fine. You do that.

Fig. 3-9. *To move a hovering helicopter to the right requires coordination of all the controls: Bell 412.*

Realize that by moving the cyclic you change the lift vector of the rotor disc so that some of the lift is now being used to move the helicopter sideways. This means that less lift is available to hold the helicopter in the air and you must increase the collective pitch slightly to compensate for this loss of lift.

When you increase collective pitch you also must remember to adjust the throttle so that main rotor rpm does not decrease and when you increase the throttle, the torque effect of the main rotor also increases, so you must counteract this torque by using the tail rotor pedals. This gets you moving to the right and everything is fine and stable until you want to stop.

Then you have to readjust each of the controls all over again.

Can you imagine what it's like to maneuver a helicopter in turbulent wind conditions? Believe me, it requires your full attention.

A few human factors are also conspiring to screw you up. Human factors are discussed in more detail in chapter 12, but two things are worth mentioning now.

Take this seemingly simple rule of thumb: "Push on the left pedal, the nose turns left. Push on the right pedal, the nose turns right."

Have you ever ridden a bicycle, sled, or soapbox derby car? Sure you have. If you want to turn one of these to the left, which hand or foot do you push forward? It's your right hand or foot. Think about it. It's a completely natural action that you have learned from experience.

I guarantee you, sometime when you're learning to fly helicopters and you're all caught up in coordinating cyclic, collective, throttle, and pedals, at least once—and probably more than once—you're going to want the nose to turn to the left and your first natural reaction will be to push the nose around with right pedal. And the nose will, of course, turn to the right. You'll note your mistake immediately and correct it by pushing on the left pedal, but your initial seat-of-the-pants reaction was to push on the right pedal. That's a human factors problem.

I'm getting ahead of myself. Chapter 12 covers human factors and safety. You haven't even had your first flight yet. That's coming up in the next chapter.

4
Your First Flight

Oh, I have slipped the surly bonds of earth
And danced the sky on laughter-silvered wings . . . .

From "High Flight,"
John Gillespie Magee Jr.

THIS SUGGESTION IS GOING TO SEEM A LITTLE RADICAL. WELL, NOT AS radical as flying into the New York TCA without contacting an air traffic controller; nothing that bad, but perhaps close.

Take your first flight in a helicopter as a passenger, not as a student pilot. (See? It wasn't that radical.)

Don't be just any passenger. Don't sit there twiddling your thumbs, reading a magazine, or enjoying the view. This is going to be part of your training. A homework assignment.

Go for a ride in a helicopter and pay attention, extremely close attention, to everything that happens. This will require some preparation and expense on your part, but I assure you the effort will be worthwhile. The preparation is to read this chapter before you go. The assignment is to observe everything possible and jot down notes that will help you remember more about the ride afterward; review the notes immediately after the flight and fill in any blanks with better explanations or observations.

WHY BE A PASSENGER?

Perhaps you're wondering "Why shell out good money just to ride as a passenger in a helicopter, when I could be using it to pay for instruction?" There are several reasons.

First, if you've never ridden in a helicopter before, how are you going to know if you like it or not? If you think you're going to like it, but aren't sure, doesn't it make sense to find out by paying a small amount for a ticket instead of three or four times as much to rent an entire helicopter and hire an instructor? What if you don't really like it?

Let me be absolutely clear about one thing: If you don't like riding in a helicopter as a passenger, you definitely should not become a helicopter pilot. Make no mistake about that.

Second, everyone's first ride in anything new usually ends up being a "gee whiz" joyride. If you do enjoy riding in helicopters—and because you've read this far in this book I assume you'll fall into this category—it's going to be awfully hard to pay attention to an instructor during the first few flights. You're going to be utterly fascinated and it will take time before the fascination wears off enough to start learning.

By taking your first ride as a passenger instead of as a student, you'll get a lot of your first "gee whiz" sensations out of the way and be ready to knuckle down to the task of learning how to fly during your first lesson: observe now, learn quicker later.

(All right. I admit you'll still be fascinated for many flights to come and you'll lose your concentration every once in awhile, but, hey, that's part of the allure of flight. At least you'll have your initial awe out of the way; more awe undoubtedly will follow.)

Third, you really can learn some things by just observing, instead of doing. Although your learning experience will be hampered somewhat by your initial enthusiasm, you'll be able to see some things that will help you later on during instruction.

Finally, if you do eventually get a civilian certificate or military wings, you'll be flying passengers someday (FIG. 4-1). They might not be as enthusiastic about helicopter flying as you are. As their pilot, you should be able to empathize with them. The best way to find out what it's like to be a passenger is to be one yourself. Seeing things from the eyes of a passenger will help you provide for their needs and make you a better pilot with passengers onboard.

Perhaps you have already ridden in a helicopter before, nevertheless I still want to urge you to follow this suggestion. It's hard to mentally retain something that doesn't make sense to you. After you've read about helicopters, you will be able to understand more and remember more.

Allow me to make an observation.

Flying in helicopters is fun, especially in nice weather. Part of the fun is seeing things on the ground from a height that doesn't make everything so small it becomes boring (FIG. 4-2). From 1,000 feet up, you can still observe people, make out familiar landscapes, and see colors. From 30,000 feet, you lose all ground detail and contrasting colors blend together—even the Rockies and Alps look dull and flat.

Fig. 4-1. *Passengers might not share your enthusiasm for flying in a helicopter: Sikorsky CH-53E Super Stallion.*

Fig. 4-2. *The view from a helicopter is much more interesting than from a high-flying airplane. You'll want to look at the sights below instead of paying attention to your instructor: MBB BK 117.*

As mentioned in chapter 1, flying lower than most fixed-wing aircraft does cause a few disadvantages, such as turbulence. But when the weather is nice and the air is smooth, you can't beat a helicopter for the view and the ride; therefore, even though this chapter reveals a few points that might seem negative, all things considered, I think you're going to enjoy flying helicopters.

That said, let's get on with your first flight.

BEFORE THE FLIGHT

One of the first things you'll notice about a helicopter is it's size. Helicopters are usually a lot smaller, especially inside the cabin, than what most experienced airplane passengers are used to (FIG. 4-3).

On the other hand, it could be just the opposite. If you ride or fly small private, commuter, or executive aircraft often and jump into a Boeing 234 Chinook that seats 44 passengers, the helicopter will not only seem bigger, it will be bigger (FIG. 4-4).

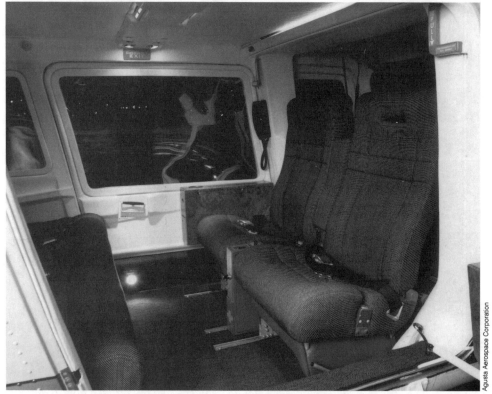

Fig. 4-3. *Helicopter cabins are usually much smaller than most airplane passengers are used to: Corporate interior of Agusta 109C.*

Fig. 4-4. *The Boeing 234 is large when compared to many commuter and charter air-planes.*

Most air travelers accustomed to DC-9s and 747s are going to find that all helicopters are smaller. Riding in the most common helicopters, like the Bell 206, Robinson R-22, and McDonnell Douglas 500, will seem more like riding in a flying sports car than in an airplane.

The largest helicopters (the Chinook, Sikorsky S-61, EH 101, and Aerospatiale AS 332, for example) if compared to airplanes, would be called medium-sized. So, you see, the label "large helicopter" is relative.

On the other hand, many people expect helicopters to be small and are quite surprised when they see the interiors of "large helicopters." I've given many tours of the S-61 and the first comment most people have is, "Oh, I didn't know it was so big inside" (FIG. 4-5). When they see the inside of the Boeing 234, they are really impressed; the commercial Chinook has the same internal finishings as Boeing's jet transports (FIG. 4-6).

Generally speaking, helicopter cabins are smaller than airplane cabins. Smaller does not mean less safe, although it does mean that some things have to be done differently. These restrictions are due solely to size—they are just as relevant in small airplanes as they are in helicopters.

Fig. 4-5. *Interior of Sikorsky S-61.*

BAGGAGE

Baggage space is usually very limited in helicopters.

In large airplanes, baggage and cargo are carried in cargo holds under the floor. In small airplanes and most helicopters, this under-the-floor space simply is not available because the cabin is not big enough for anything under the floor. In medium helicopters, the fuel tanks are under the floor. (Airplanes have standard fuel tanks in the

Fig. 4-6. *Interior of Boeing 234.*

wings; auxiliary tanks might be located in a baggage compartment.) The only helicopter with under-the-floor cargo holds is the S-61 and even these compartments are not very big (FIG. 4-7).

 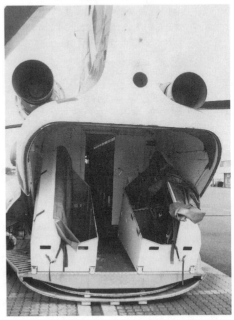

Fig. 4-7. *Baggage compartments in (A) Aerospatiale AS 332, (B) Boeing 234,*

Fig. 4-7, continued.
(C) Enstrom F-28F, and

(D) Bell 212.

Where does the baggage go then? Often, there are small cargo holds in the fuse-lage, usually behind the cabin. The only baggage compartments in the little Robinson R-22 are under the seat cushions. Sometimes a helicopter operator finds it necessary to block off a portion of the cabin itself and put the baggage in there. This is safe and legal if the baggage is secured properly.

Unfortunately, most of the large empty area in a helicopter's tailboom, that part of the fuselage between the cabin and the tail rotor, must be kept empty. This is because too much weight in the back will move the center of gravity of the helicopter too far aft, which will make the aircraft tail heavy.

Carry-on baggage space is also very limited. Most helicopters, unless they are configured for executive travel, don't have facilities for hanging bags. Even if you take a small bag or briefcase inside the cabin with you, you might end up holding it on your lap. In some small helicopters, the passenger seats are mounted right on top of specially-built fuel tanks and there is no room to put anything under the seats. This sounds hazardous, but is no more dangerous than the fuel tank in the trunk of your car. Actually, it is probably safer because aircraft have stricter construction requirements than automobiles.

The main point to remember about baggage is that there isn't a lot of room in a helicopter. I'm not going to tell you not to take that big suitcase with retractable wheels and 50 pounds of carry-on luggage—after all, if you've just crossed the Atlantic and want to take the helicopter shuttle from Kennedy to downtown Manhattan, you'll probably have a lot of luggage. This is one of the things the helicopter operator must consider when he chooses the type of helicopter he will use for a certain operation and how many passengers it can carry.

Otherwise, please, if you're just going to make a short trip that includes a flight in a helicopter, at least consider packing as lightly as possible. If you must carry a lot of luggage, divide it up into two or three small bags. From the pilot's point of view (the pilot often being the baggage handler), two medium-size soft-sided bags are preferable to one big hard-sided suitcase because two soft-sides are easier to lift and stow. Try to limit carry-on luggage to a camera bag, briefcase, or purse. You and the other passengers will have a lot more leg room.

CLOTHING

What should you wear during the flight?

I have the feeling that very few passengers think very much about the clothes they should wear when traveling by airplane or helicopter. Of course, in many places in the world oil workers are required to don antiexposure survival suits for their helicopter flights to and from their offshore destinations (FIG. 4-8). But for the most part, airplane passengers probably don't even think about wearing special clothing when they travel by air.

The airlines are regrettably closed-mouthed about the subject. After all, they want people to believe that air travel is as safe as ground travel, safer actually. And statistics generally prove they're correct. The airlines are apparently reluctant to talk about wearing special clothing on aircraft because that would make aircraft seem less safe. It might also scare some passengers away and that's not good for business. The airlines want people to want to fly.

So do I, but I think a frank discussion about clothing is worth losing a few potential passengers if it can help some real passengers avoid a good deal of pain and suffering—and perhaps save a life or two.

Not much effort nor deviation from current fashion styles is necessary to clothe yourself properly when traveling by air.

Fig. 4-8. *Passengers on offshore helicopter flights in some parts of the world are required to wear antiexposure suits: Helikopter Service Boeing 234, North Sea.*

Aircraft accidents are rare, but they do happen. What if you have to evacuate from an airplane or helicopter quickly? What if there is a fire on board? What if there is snow outside? Or what if something as mundane as fog causes your flight to divert to another airport with colder weather than your expected destination? These are things you should think about when choosing clothes before a flight.

Fashion being what it is, the passenger dressed in a comfortable wool business suit is better dressed for an emergency evacuation than a fashion-conscious passenger wearing lightweight clothing, for instance a man wearing a short-sleeve sport shirt, shorts, and sockless shoes, or a woman wearing a dress or skirt, a thin blouse, high-heeled shoes, and hose (FIG. 4-9).

The better options for all passengers are nonsynthetic slacks, shirts, blouses, jackets, and flat shoes with cotton socks.

Women should avoid hose because in case of a fire or intense heat, nylon has a tendency to melt into skin as it burns. Natural fabrics burn, too, but at least they don't melt. Wool is the best, and the more layers of clothing between your skin and the fire the better.

If you really want to get down to the nitty-gritty, a high-necked shirt or sweater will protect your neck, a hat will keep your hair from burning, and gloves will protect

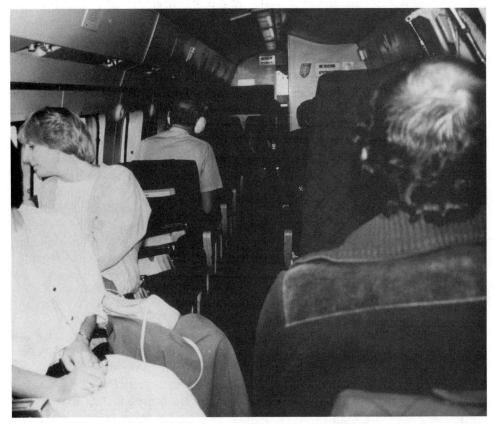

Fig. 4-9. *Most airplane and helicopter passengers give little thought to the protective function of their clothing in case of a diversion to another airport, a precautionary landing, or an accident.*

your hands. You don't have to sit for the whole flight wearing your gloves and hat, having them handy in carry-on baggage is fine. It only takes a few seconds to put them on and you'll probably have at least that much warning in the event of an emergency.

When I ride as a passenger, I wear my jacket or coat during takeoffs and landings. Statistically, it is during these two phases of the flight that most emergencies occur. If we must evacuate quickly, I have protected myself as best I can against fire on my way out of the aircraft and against the weather once I get outside.

Point: The number of emergency exits on aircraft is determined by how fast a full load of passengers can evacuate in an emergency. The aircraft manufacturer must prove with actual tests that all the passengers can get out within a certain time period.

Counterpoint: These tests are not conducted with an average passenger mix of children, adults, and persons who are old or handicapped. Tests are done with physically-fit, young adults who are wearing jogging suits and running shoes, who practice until they can get the evacuation done in the minimum required time.

Clothing is going to protect you while you wait to get through that emergency exit and it might save you from serious injury, perhaps even fatal injury.

Did you ever think what it will be like inside the cabin if you have a rapid decompression at 30,000 feet? It's going to be cold, as low as -60°F or lower, even in the summer. You'll get an oxygen mask and the pilots will begin a fast descent, but it will take some time to get down to warmer temperatures. Wouldn't you rather to be wearing a long sleeve shirt and a sweater than just a T-shirt or blouse? Wouldn't it be nice to have a jacket within easy reach?

Helicopters aren't pressurized and don't fly high enough to require it, but they can be cold and drafty. What if the heater fails or doesn't work? What if the pilot has to land out in the wilderness because of a mechanical problem? Are you dressed for a hike in rough terrain?

The old Boy Scout motto is a cliché, but it wouldn't hurt if more people took it to heart when they traveled by air. The clothes you should wear in a helicopter really aren't any different than what you should wear in an airplane. With a little thought, you can "Be Prepared" for the worst, and no one will even notice you've disregarded the fashion designers for the duration of your flight.

HEARING PROTECTION

Although improvements are being made all the time, most helicopters are still noisier inside than airplanes. The noise comes from the engines, which in many passenger-carrying helicopters are turbine engines, from the transmission gears, from the rotor blades, and from the passage of the fuselage through the air.

Airplane designers can locate engines away from the cabin and they don't have to worry about the noise from a transmission or rotor blades. Wind noise is dampened out by sound insulation surrounding the pressurized fuselage. The ride is very quiet.

Helicopter designers have a much harder job. The Boeing Vertol 234, for example, is about as quiet (or noisy) as a turboprop airplane. Unfortunately, sound insulation means extra weight and operators often opt for less insulation and more available payload. Insulation doesn't pay its way (FIG. 4-10).

Unless you happen to ride an extremely well-insulated helicopter, you won't be able to talk easily to the person sitting next to you. The use of hearing protection is necessary to protect your hearing and also tends to make the flight less fatiguing.

Some operators equip their machines with in-flight entertainment systems. These systems are usually connected to individual headphones for each passenger. The headphones not only provide you with music and announcements from the crew, but also dampen out the noise.

If the helicopter is not equipped with an in-flight entertainment system, the operator will usually provide passengers with plain earmuff headsets or inexpensive disposable earplugs. Earplugs aren't as comfortable as headsets, but do take out the worst part of the noise.

If you have your own earmuffs or earplugs, you can certainly use them on the helicopter. Industrial earmuffs are the most effective.

Fig. 4-10. *British Caledonian Helicopters installed extra sound insulation in the Sikorsky S-61s they used on the shuttle between Heathrow and Gatwick airports in England so that their passengers wouldn't need hearing protection.*

The spongy earphones that come with portable cassette players won't remove much noise, but if you use earplugs and turn up the volume high enough, you can enjoy your own choice of taped music. (Earplugs will filter out the high frequency noise of the engines and transmission; even though you'll have to increase the volume of the music, you'll still have a net reduction in the total noise level.) Don't count on using an AM/FM radio, however. There will probably be too much static interference in the helicopter to get good reception.

Some airlines prohibit the use of portable radios in flight because of interference with the aircraft's navigation equipment. I have never heard of this restriction on helicopters. Helicopter navigation equipment has to cope with so much inherent interference from the aircraft (because of all the rotating parts), that a small portable radio receiver probably wouldn't make much difference. But it would be smart to ask permission prior to engine start or takeoff.

TOILET FACILITIES

Most helicopters, even the larger ones, do not have toilet facilities on board. The exceptions are the Boeing Vertol 234 and selected Sikorsky S-61s that have chemical toilets. Donald Trump's private Super Puma had a chemical toilet, too, but it was crammed into such a small place that it probably was never used. Fortunately, most helicopter flights are not very lengthy, but the lack of facilities is something to consider.

It's more than something to consider. Most pilots are avid coffee drinkers, but they are careful about how much they drink before and during a flight and the last place they stop before going out to the helicopter is usually the restroom. So take a tip from pilots based upon experience.

In a dire emergency, you might find that a helicopter is equipped with special bottles in lieu of toilets. Some of these are cleverly called H.E.R.E. bottles, for "Human Endurance Range Extenders." But don't count on the fact that the helicopter has any. If it were an extreme emergency, I guess you could ask the pilot to land in a field somewhere, but this could be very embarrassing. Use the restroom before you board.

Will you become air sick in a helicopter?

That's a hard question to answer because it depends so much on the individual.

I suspect that all pilots have become sick or at least felt queasy at some time or another while flying. Experienced pilots rarely feel queasy in flight because their bodies are used to the motion of the aircraft.

Some air sickness is caused by just plain nervousness, whatever its cause. On my first training flight in a Cessna 172 in the Air Force, I got sick and thought my whole flying career was at an end; however, the flight surgeon assured me it was probably because I was nervous about it being the first flight. I didn't believe him because I didn't feel nervous, but I hoped he was right. To my pleasant surprise, I didn't get sick on the next flight nor the next nor the next and I haven't been sick since, although I have felt queasy a few times for various reasons.

If you know you have a propensity toward motion sickness, you can take a few precautions before the flight. A stomach full of greasy food obviously isn't going to help things, but a completely empty stomach can be just as bad. Hunger pangs sometimes evolve into motion sickness. Eat a small amount of something bland before the flight.

Carbonated beverages help some people and hinder others. They do make everyone burp more. The higher the aircraft goes, the more you will burp because air pressure decreases and the gases trapped in your body force their way out. If burping tends to aggravate your motion sickness, then it's probably best to avoid carbonated beverages.

Alcohol might reduce your nervousness and eliminate that one cause of motion sickness, but it might create another by upsetting your stomach. Too much alcohol, a big meal, and a turbulent flight make motion sickness almost a sure thing.

Numerous medications help reduce motion sickness. If over-the-counter medicines don't help, your doctor can prescribe stronger ones. A very good one called scopolamine was originally designed for NASA astronauts in zero-G environments (weightlessness). It comes in a small circular adhesive bandage that is placed on the neck behind the ear and is good for 72 hours. The medicine is absorbed through the skin and does not cause drowsiness in most people.

If you find yourself becoming queasy in flight despite all these precautions, there are a few things you can do that will hopefully calm down your stomach.

First, remember that motion sickness is related to your sense of balance, which is felt in your inner ear. When you feel queasy, it's because there's a discrepancy between what your inner ear is feeling and what your eyes are seeing. If you are reading or looking only inside the aircraft, your eyes will tell your brain you are sitting upright, but your inner ear will be detecting all the movements of the aircraft.

Look out the window and give your eyes and inner ear a chance to agree (FIG. 4-11). Don't look down, rather straight out toward the horizon. Take slow and steady deep breaths. Don't read and don't make sudden head movements that will also upset the balance in your inner ear. Don't eat—you probably won't feel like eating anyway—and don't drink anything stronger than water and only in small quantities.

If you still become sick, there should be a motion sickness bag nearby you. Please use it and don't be embarrassed. It can happen to everyone and probably has.

Fig. 4-11. *Motion sickness is often caused when your sense of balance doesn't agree with what your eyes are seeing. By looking outside the aircraft, you help your senses of sight and balance get in sync.*

BOARDING THE HELICOPTER

Depending on circumstances, you may board the helicopter while it is shut down, while it has one or more engines running, or while the rotors are turning (FIG. 4-12). All ways are safe, but with all three you must take certain precautions.

When the helicopter is shut down, things are quiet and nothing is moving. Everything looks safe, but make sure the cockpit crew is not busy with engine start-up procedures.

The most dangerous time to be near a helicopter is when the rotors are starting or stopping.

When the rotors are stopped, special devices called *droop stops* keep the blades from hanging down too far. When the rotors are rotating at normal speed, the rotors can droop down quite low, but the pilot has full control of their position with the cyclic and collective sticks. During start-up, after the droop stops move out of position but before the rotor blades are up to normal rpm, the blades are not moving fast enough to be fully controllable by the pilot and are therefore very susceptible to wind gusts.

A gust of wind at the wrong instant can cause a main rotor blade to flap down so low that it can hit the top of the cockpit or tailboom. This is the main reason why helicopters have wind limitations for start-up and shutdown. Needless to say, a blade could also flap down low enough to hit a person standing within the circumference of the rotor disc. It has happened.

How can you tell if a helicopter is starting up? If there is no one sitting in the cockpit, you're safe. The helicopter cannot start itself. If there is someone sitting there, you should assume he or she might be starting the engine(s); therefore, approach the helicopter from the front. Stop outside the tip of the rotor blade and get a clear signal from the person sitting in the helicopter before moving any closer (FIG. 4-13).

I should mention that the person in the cockpit does not necessarily have to be a pilot. Certain aircraft mechanics are authorized to start the engines and engage the rotor blades, too. So, simply because a person in the cockpit is not wearing a pilot uniform, don't assume that they cannot start the rotors.

Engine noise is unmistakable, but not necessarily an indication that the rotors will soon be turning. Most twin-engine and many single-engine helicopters are equipped with a *rotor brake* (very similar to a disc brake on a car) that is used to stop the rotor after the engines are shut down. Selected helicopters are started with the rotor brake engaged and, as a consequence, you'll hear one or both engines, but the rotor will not be turning. Other helicopters are started with the rotor brake off; in this case, the rotors will start turning at the same time the first engine is started.

To summarize, the rotors can't turn without at least one engine running, but a running engine does not necessarily mean the rotors will be turning.

Approaching a helicopter from the front is also the best practice when the engines are running and when the rotors are turning, too. In both cases, someone will be sit-

Fig. 4-12. *These passengers are boarding the Sikorsky S-61 while the rotors are turning. They were escorted to the helicopter by ground personnel.*

Fig. 4-13. *A helicopter cannot start by itself, but if there is someone in the cockpit, the engines might start soon. Approach from the front and don't walk under the tip path plane of the main rotor blades until you get a clear signal from someone that is aware of aircraft operations: MBB BK 117.*

ting in the cockpit. If you are not being escorted to the aircraft by ground personnel (who presumably receive clearance from the cockpit crew to approach the helicopter) then be sure you get clearance from the cockpit crew before moving under the rotor disc.

How do you get clearance to move in toward the helicopter? Wait in front of the cockpit until you get the pilots' attention. Don't bother to shout because they can't hear you. If they're looking down at something inside the cockpit, wait until they look up. You won't have to wait long.

They'll either wave you in or give you the universal thumbs-up signal. Pilots like to use thumbs-up because it looks cool.

Never, never, never approach a helicopter from the rear.

Tail rotors spin fast, are hard to see, are often low to the ground, and will kill you in an instant.

Even if you stay clear of the tail rotor, the pilots are not going to know you are there. If they don't know you're there and they decide to take off at the precise moment that you pass the tail rotor, you're going to have a problem.

Another reason not to approach a helicopter from the rear is the engine exhaust. Exhaust gas temperatures might be as high as 600 °C. Of course, the gases cool off quickly when they hit the air and the farther away from the exhaust ducts you are, the cooler the gases, but they still can be warm enough to be uncomfortable.

To tell the truth, on very cold days I have stood in the exhaust in order to stay warm, although not for very long. The gas odor is strong and your hair gets dirty.

BEFORE TAKEOFF SAFETY BRIEFING

Small cabins mean no cabin attendants, stewards, or stewardesses. This is legal up to 10 passengers; operators are often able to get a dispensation that allows them to fly with up to 19 passengers or more without a cabin attendant.

This means that you the passenger are going to be more on your own, more responsible for your actions and safety. Remember this fact when you listen to the passenger briefing (FIG. 4-14).

Preflight passenger safety briefings are required by most aviation authorities and in the absence of a cabin attendant, one of the pilots will usually give this briefing. Or, you might get the briefing prior to boarding either by video or from a ground attendant.

Everyone who flies frequently has heard passenger safety briefings on aircraft. They also know that most passengers don't pay attention to the briefings. Unfortunately, the noise level inside many helicopters might make the briefing almost unintelligible, even if a public address system is used. The net effect is that many passengers don't know what was covered in the briefing.

Forgive me if I sound like I'm preaching here, but the passenger safety briefing really is important to you, particularly in an emergency. Without a cabin attendant on board in a helicopter or an airplane, the passengers really do have to look out for themselves because the pilots are going to be very busy in the cockpit. So, if you can't hear the briefing or miss something in it, look for the briefing card or folder that is required by law to be available for each passenger and read it carefully. Sermon over.

Of course, sometimes even the best-intentioned briefings go astray. Misunderstandings because of language can cause problems with interesting results.

Such an incident occurred on a Helikopter Service flight. A Norwegian helicopter captain was briefing a group of Italian passengers in English. He discovered very soon that few of them understood what he was saying, so he asked if there was anyone who could interpret for him. One man stood up right away and said, "Yes, yes, I speak English." The captain gave most of the normal briefing, pausing often to allow his interpreter to repeat the instructions about emergency exits, seat belts, and other pertinent items.

Fig. 4-14. *Emergency exits and safety equipment are two important items covered in passenger safety briefings: (left to right) fire extinguisher, life raft mounted in emergency exit, emergency locator beacon, flashlight, first aid kit, and (lower right corner) life vests: Sikorsky S-61.*

When he came to the subject of life vests (which were required to be worn during this offshore flight), he demonstrated how to put the vest on. His interpreter did the same thing and the other 17 passengers followed suit. Then the captain explained how one should pull down on the red handles to inflate the vests in the event that the helicopter had to land on the water. He waited patiently while the interpreter explained this in Italian, but to his horror, the next sound he heard was the whoosh of 36 bottles of pressurized carbon dioxide inflating the 18 vests of his passengers.

SEAT BELTS

Once in the aircraft, be sure to fasten your seat belt. A pilot or ground attendant might check that all the passengers have fastened their belts, but it's better to take care of yourself (FIG. 4-15). Observe and obey the seat belt signs.

Size puts a restriction on movement inside the cabin. In the small helicopters, there is about as much room as a car. You can't get up and walk around inside your car while driving, can you?

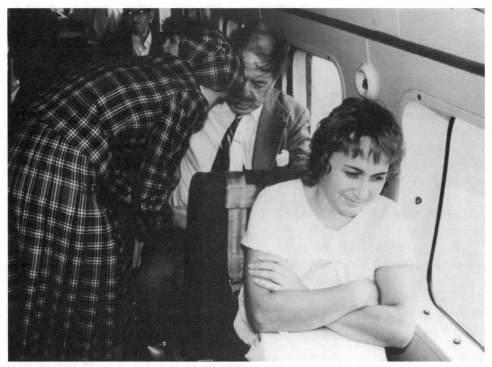

Fig. 4-15. *British Caledonia Helicopters ground hostess checks the passengers' seat belts.*

Large helicopters have some room to move around, especially if the aircraft isn't full, but there really isn't anywhere to move to. For passenger safety, the fasten seat belt sign is usually left on during the entire flight. Without a cabin attendant in the back, most pilots are reluctant to allow passengers to move freely around the cabin, but they do make exceptions if you have a special request.

Besides, with no toilet facilities on-board, there really isn't any need to move around.

SMOKING

Another problem with small aircraft cabins is the lack of smoking and nonsmoking sections. There just isn't any sense in making separate sections when everyone is sitting shoulder to shoulder.

Even in large helicopters, dividing up smokers and nonsmokers really does not make much difference to the nonsmokers. They end up breathing the noxious fumes anyway.

What can you do? Some companies that own or lease helicopters often have arbitrarily decided that there will be no smoking on any of their flights. Others have arrived at the same decision by democratic vote. Of course, the pilot-in-command always has the option to refuse to allow smoking on the aircraft, but most pilots and helicopter operators with a concern for passenger service would rather defer the decision to the customer or individual passengers.

If you know who your fellow passengers will be, perhaps you can come to some agreement about smoking before the flight. Or, if for some reason, perhaps medical, you really can't tolerate smoke, it wouldn't hurt to mention this to the pilot (and the other passengers.) The point is you'll be in a confined space that unfortunately often does not have the best ventilation. Most smokers are willing to abstain for a while if they know their smoking will cause problems for other people.

Smokers' needs should be considered because smoking has a calming effect on many people. I flew survivors of a tragic North Sea accident that claimed 123 lives and I have never seen so much smoke in a cabin. I think even the nonsmokers were smoking.

Or, when confronted with a smoker, you could hand him a copy of this book. It might help calm his nerves enough so that he doesn't need to light up.

Smoking is not allowed during takeoff and landing, the obvious reason being the fire hazard in case of an emergency. Neither is smoking allowed on the ground near the aircraft. If you do want to light up at an airport or heliport, do so only in a designated smoking area.

SITTING NEXT TO THE PILOT

The trend in passenger-carrying operations is toward a two-pilot crew in a twin-engine helicopter, especially at night and in poor visibility conditions, called *instrument meteorological conditions* (IMC), when flying must be done by reference to cockpit instruments only and in accordance with instrument flight rules (IFR). Vast improvements in autopilot systems for helicopters have made it possible for certain helicopter manufacturers to obtain approval for single-pilot IFR operation under specific conditions. Many helicopters are approved for operation by a single pilot when the visibility is good and visual flight rules (VFR) are in effect.

Whatever the case, you might find yourself sitting in the left-side seat of the cockpit, next to the pilot; a helicopter copilot would occupy this seat when required. (This is exactly the opposite of the seating arrangement in an airplane cockpit. The discussion about how this came about has been going on in helicopter publications for years.)

Regulations might require that the copilot's controls be removed before a passenger may ride along in the copilot's seat. If the controls are there, don't touch them. And don't touch any of the switches, knobs, or other cockpit apparatus.

If the pilot has time, he will probably be glad to explain to you how some of the systems work. He might even give you a chance to take the controls and fly a little; however, please don't bug him about this. If he's busy or stressed or just plain tired, he might not want to take on the extra burden of teaching a passenger how to fly. Just sit back, enjoy the ride, and be grateful that you have a chance to sit in one of the best seats in the house.

THE FLIGHT

Let's assume that you board the helicopter while everything is quiet; you find a seat by a window because helicopters have lots of windows, put a small, carry-on bag under the seat, and fasten your seat belt. You listen carefully while one of the pilots gives the passenger briefing. Soon all the luggage is loaded, the doors are closed, and the pilots take their places in the cockpit. It's time to start the engine(s).

You notice that the pilots wear headsets. This is so they can converse easily with each other and over the radio. If you are at an airport or heliport with air traffic control, the pilots will probably have to get clearance before they may start the engines.

Start-up

As soon as the pilot presses the start switch, you'll hear the starter begin to turn one of the engines. The noise gets louder and louder as the engine rotates faster and faster. Variation in the sound of the engine is likely as the pilot or an automated system controls the amount of fuel being injected into the engine. (The pilot must avoid a *hot start*, which is the result of too much fuel and not enough air.) Once the rotation of the engine increases above a certain rpm, the engine is self-sustaining and the pilot releases the starter, or it disengages automatically.

The rotor blades might begin turning when the first or only engine is started, depending upon the type of helicopter, the approved starting procedure, or the wind. Certain helicopters are started with the rotor brake off in a no-wind condition and rotor brake on in high wind. Leaving the rotor brake on until one engine is up to a prescribed power setting will allow the rotors to accelerate faster when the rotor brake is released. A faster rotor acceleration is an advantage in high winds because the rotors become controllable sooner.

When the rotor blades start to turn, the entire fuselage of the helicopter will begin to sway back and forth. As the rotors go faster and faster, the swaying will stop, but you'll feel a fast (high frequency) vibration throughout the cabin.

I remember riding in an old Air Force helicopter, the H-43 Huskie, that had two counter-rotating main rotor blades and no tail rotor. The Huskie shook so much during start-up that I thought it was going to fall apart, but once the rotors came up to normal rotational speed, the shaking diminished. Fortunately, modern helicopters don't shake nearly as much as the H-43 when starting up.

Pilots prefer to start up with the nose of the helicopter facing into the prevailing wind. Sometimes this is not convenient and it is perfectly safe to start up with a slight crosswind or tailwind. The only problem is that the combination of the wind and the downwash from the rotor blades might cause some of the exhaust from the engines to circulate into the cabin. This can be quite uncomfortable, fortunately it usually takes only a few moments until the rotors come up to full rpm and the exhaust is blown away from the helicopter.

Pilots try to make sure this doesn't happen, but sometimes they're unlucky. One time, while starting up on a hot, windless day, I purposely left the cabin doors open so the cabin wouldn't become too warm for the passengers who were wearing antiexposure suits. (The cabin ventilation fan was powered by the AC generators that would not produce electrical current until the rotors and main gearbox were turning.)

I didn't realize there was a very slight tailwind and as the engines started, the exhaust was sucked directly into the cabin. By the time the other pilot and I smelled the exhaust in the cockpit and were able to signal the mechanic to close the cabin doors, all the passengers were teary-eyed and coughing. I felt very bad about what happened, but all I could do was apologize and explain the situation as best I could over the PA system.

Taxiing

Once the rotors are turning and everything else is turned on and checked, the pilots taxi the helicopter to the takeoff position. If the helicopter doesn't have wheels, then it has to be *hover taxied*—flown. If the helicopter has wheels, the pilots will usually taxi normally to conserve fuel (FIG. 4-16). Abnormal circumstances, such as ice or snow on the ground, might make it necessary to hover taxi a wheeled helicopter.

The time from engine start to hover taxi will probably be a bit longer than from engine start to ground taxi because more checks must be performed before a helicopter is lifted into a hover.

Fig. 4-16. *Helicopters with wheeled-landing gear are normally ground taxied to the take-off position.*

In fact, every system in the helicopter is checked either before or after the engines are started. Depending upon the type of helicopter and company procedures, the pilots might do a power assurance check of the engine or engines before lifting into a hover. You might hear changes in the engine noise as each engine is checked at a specified power setting. The pilots must also check the operation of the freewheeling gears and this might cause some strange sounds as each engine disengages and engages with the main gearbox.

Takeoff

Although helicopters, both those with wheels and skids, can take off from a runway like an airplane, most of the time they take off from a hover. An important point to remember is that a hovering helicopter is a flying helicopter.

In a way, this makes a helicopter safer than an airplane. An airplane is not flying until it has reached a certain minimum airspeed. To come up to its minimum flying speed, an airplane must accelerate along the ground, usually using full power from its engines to make the takeoff roll as short as possible.

This is a critical time for an airplane—any emergency must be dealt with quickly and correctly. And because you can't fully check the health of some things while stationary on the ground, like the full power output of the engines and the effectiveness of the controls, there's a period of uncertainty until the airplane has attained flying speed, lifted from the ground, and flown to a safe altitude.

A helicopter cockpit crew can check all of these things in a hover (FIG. 4-17). Because a helicopter uses more power in a hover than in cruise flight, the pilots know that if the machine has enough power to hover, it has more than enough power to fly straight and level. Hovering also requires more control inputs than in cruise flight; again, if the pilot can control the helicopter in a hover, he knows that it can be controlled in flight as well.

If something is not working properly, it's very easy to put the helicopter back down on the ground. The pilots might discover a discrepancy while lifting into the hover and might never leave the ground at all. So, the period of uncertainty—Will it work this time or not?—is much shorter in a helicopter than in an airplane. The period actually lasts only the few seconds it takes the pilot to lift the helicopter from the ground into a stabilized hover.

What happens after the hover depends on many factors. One thing is certain, most of the time the helicopter will not go straight up like an elevator.

This misconception, that helicopters take off straight up, is almost common enough to be a myth. Most people who have ridden helicopters or seen them take off know this isn't so, but those who have only seen helicopters on television or in the movies seem to think that it is accurate.

A friend in the Air Force used to do what he called an "FBI takeoff." He said he had picked it up from watching TV shows, mainly the old "FBI" series.

Fig. 4-17. *During the predeparture hover check, the pilots check the engines, flight controls, and other critical systems before transitioning to forward flight: Japanese Defense Force AS 332 Super Puma.*

He would lift the helicopter into a hover, climb straight up to 100 feet, kick in a 180-degree turn with the pedals, and then push the nose over to gain airspeed diving toward the ground. He could only do this takeoff when the helicopter was very light (and therefore had a lot of power available). It was a blatantly unauthorized maneuver performed with no passengers and no witnesses on the ground.

A vertical takeoff, one that goes straight up, requires an enormous amount of power compared to cruise flight or even a low hover over the ground. Even when a helicopter has the power available to make a vertical takeoff, it is a waste of fuel and places extra strain on the engine and other parts of the machine.

To some extent, it is also unsafe. All helicopters must operate in accordance with certain procedures that are determined by performance capabilities. Certain altitude and airspeed combinations negate a safe landing if the engine fails. It's the pilot's job to always avoid these unsafe combinations.

Forget vertical takeoffs. The only time a pilot will do such a takeoff is when obstacles must be avoided. For example, an air ambulance pilot might have to land close to a patient in an area surrounded by trees or telephone wires. He knows the risks involved and tries to give himself as much power as possible by reducing the helicopter's weight, primarily by limiting fuel and passengers.

How does the helicopter take off if it doesn't go straight up?

It depends on a lot of factors, including helicopter type, number of engines, wind, temperature, visibility, air pressure, length of the runway or helipad, obstacles, and local restrictions or regulations, such as noise abatement procedures, but basically this is what happens.

Let's assume we're taking off from a flat surface, on the ground, and clear of obstacles: airport or heliport runway, or heliport with an open area under the takeoff path.

To start the helicopter moving forward, the pilot moves the cyclic stick in his right hand slightly forward. This causes the nose of the helicopter to pitch down, which results in simultaneous forward and downward movement (FIG. 4-18). To counteract the slight descent, the pilot increases power with the collective lever in his left hand. The net result is that the machine begins to move forward at a faster and faster rate while staying level with the ground.

Fig. 4-18. *The amount of nose-down attitude required during takeoff varies with helicopter type and conditions: Helikopter Service Sikorsky S-61N departing Forus Heliport.*

This nose-down movement seems exactly opposite to what an airplane does during takeoff, but it isn't. An airplane must take off with forward airspeed and, in a sense, trades some of this forward speed for vertical speed at the point it leaves the ground; when the required takeoff speed is obtained, the pilot pulls the nose up to cause the airplane to climb. What you don't see when an airplane takes off is the pilot holding the nose of the airplane down with forward pressure on the control stick or column during the takeoff roll until the craft reaches the required takeoff speed.

The helicopter begins a "takeoff roll" in a hover with zero airspeed (depending on the wind, of course); therefore, in order to reach the most efficient climb speed, the pilot must also get the helicopter moving forward first. The difference is that the airplane is accelerating nose down on the ground and the helicopter is accelerating nose down a few feet over the ground.

At about 15 to 35 knots, the helicopter begins to benefit from *translational lift*, which is the additional lift obtained through airspeed because of the increased efficiency of the rotor system. This means that while it might take 90 percent of a helicopter's available power to fly level at 20 knots, it might take only 80 percent to fly at 45 knots, and 65 percent to fly at 120 knots.

You might feel a slight burble of turbulence as the helicopter begins to encounter translational lift; it's nothing to worry about. Translational lift is an aerodynamic law that can't be repealed. Helicopter type, wind, pilot technique, and other factors can make the passage into translational lift more or less noticeable.

The pilot allows the helicopter to accelerate to climb speed, usually between 80 and 100 knots, and then eases back on the cyclic to hold that climb speed. Depending on the air route structure, other traffic, and obstacles, it might be necessary to level off and make a few turns before climbing to cruising altitude.

Shortly after takeoff, the pilots will do the after takeoff checklist, which varies among helicopter types and flight departments. If the helicopter has retractable landing gear, the pilots will raise the wheels and you might hear the whine of a hydraulic pump as the wheels come up and a dull clunk when they're in place. The only other after takeoff checklist item you might notice is the no smoking sign going off.

Cruise

In cruise flight, a helicopter is not much different than an airplane. It might seem slower than usual to you. It is. It might seem lower than usual to you. It is. It will probably be noisier than you're used to. It is. Other than those things, it's not much different.

One unusual noise you might hear is *blade slap*, a very loud WHOP-WHOP-WHOP that is a characteristic of helicopters with two main rotor blades (FIG. 4-19). You'll hear it often when the helicopter is turning fairly steeply and it's even louder on the ground. For this reason, pilots try to avoid blade slap, but because it is dependent upon airspeed, temperature, and pressure altitude, sometimes the slap sneaks up and before you know it, there it is.

Bell Helicopter Textron

Fig. 4-19. *Blade slap is common to helicopters with two main rotor blades, such as this Bell 206L LongRanger. With proper pilot technique, blade slap can be avoided under most atmospheric conditions.*

The view from a helicopter on a clear day is beautiful no matter where you are. It doesn't matter if you are over a city or countryside or wilderness or water, although after a while water can get downright monotonous. Everywhere else, interesting things are always available to see and photograph, if you like to do that.

Meal and beverage service on helicopters tends to be primitive, if not completely nonexistent. You'll be lucky if you get a bottle of coffee and some cookies to share with the other passengers. Without a cabin attendant on board, self-service is in order.

Plan on eating and drinking before and after the flight or bring your own. Check before you open that bottle of whiskey because consumption of alcoholic beverages might be frowned upon by company, customer, or aviation regulations.

The lack of cabin attendants on most helicopters causes another problem: How do you give a message to the pilots?

The surest way is simply to write your message on a piece of paper and hand it to one of the pilots. Try to hand it to them when they are not busy flying or talking on the radio. Don't worry if the seat belt sign is on because you may leave your seat to give the pilots a message anytime except during takeoff and landing.

The worst way to tell the pilots something is to shout the message to them. The noise is so loud that you'll probably have to repeat the message several times and the pilots might ask you to write it down anyway to be sure they understand you.

The best way to talk to the pilots is through a microphone on a headset. If there is an extra headset available, the pilots will probably ask you to use it if your message is long. In most machines, you'll have to push a switch to talk and release it to hear the reply. Certain aircraft intercoms are *voice-activated*, which means you don't have to push a switch to talk, although you might have to talk slightly louder than normal to get the system to work. If you can hear yourself talking in your own headset, then the other people with headsets probably hear you, too.

Landing

Vertical landings, like vertical takeoffs, are very unusual. An elevator-like straight-down landing is even more unusual than an elevator-like straight-up takeoff. It's also more dangerous.

The danger is due to settling-with-power, which was explained in chapter 1.

Helicopter pilots avoid settling-with-power like the plague. Vertical landings are possible, but a good margin of power is necessary so that the descent can be done very slowly and the possibility of entering settling-with-power is minimized.

A standard landing in a helicopter is very similar to a landing in an airplane, except the helicopter usually stops in a hover with zero groundspeed instead of squeaking down on the runway with 60 to 100 knots or so. Helicopters can also land with some forward speed and many operators have their pilots do this routinely; however, because the landing gear in helicopters is not built for high speeds, forward landings are usually limited to no more than 40 knots. Helicopters with skids can also land with forward speeds, but this is done only during certain emergencies.

Before landing, the pilots do a prelanding checklist. As with the after takeoff checklist, you might hear the landing gear being pumped down and locked into position. The no smoking light will come on again and a pilot might give a short passenger briefing.

Landing is a busy time for the pilots (FIG. 4-20). Even in visual conditions, there are procedures to be followed, radio calls to make, and often other traffic to avoid. When in the clouds and flying an instrument approach procedure, the pilots are even more busy; therefore, it is important not to bother them with messages that don't have an impact on the safety of the flight.

You'll hear changes in the engine noise as the pilots adjust the engines in the descent and as the helicopter nears the ground. Unlike airplanes that land with low-power throttle settings, a helicopter landing to a hover will need a high power setting because hovering takes more power than cruise flight. You'll first hear decreasing engine noise as the helicopter descends toward the touchdown area and then increasing noise during the transition into the hover.

You're also likely to feel an increase in vibrations as the helicopter passes out of translational lift and into a hover (FIG. 4-21). The S-61, for example, is particularly susceptible to vibrations during this period and it takes very precise control inputs to avoid what can seem like very alarming vibrations. On the other hand, the AS-332 Super Puma experiences almost no vibrations when landing. Like many things in flying, changes in temperature, pressure altitude, wind, and the weight of the helicopter can make differences in the vibration level during landings.

Fig. 4-20. *Landing is one of the busiest times for the cockpit crew of any aircraft, particularly at busy airports: Sikorsky S-61, Heathrow Airport.*

After stopping in a hover over the landing spot, the pilots will taxi to the assigned parking place. In a helicopter with skids, this will require a hover taxi. In one with wheels, the helicopter will normally be ground taxied. Please remain in your seat with your seat belt fastened in both cases. Ninety-nine times out of one hundred, nothing will happen, but there's always the chance that the pilot might have to make a quick movement on the controls to avoid an obstacle or react to an emergency and then it's best to be seated and strapped in.

After Landing

Circumstances might require you to disembark from the helicopter when the rotors are still turning (FIG. 4-22). Use the same precautions you did when boarding the aircraft: Follow the instructions of the ground personnel and stay away from the tail rotor. Although it is a very dangerous time to be under the rotor disc when the

Fig. 4-21. *As a helicopter decelerates through translational lift, you'll feel an increase in vibrations: Bell 206 JetRanger.*

Fig. 4-22. *Always stay away from the tail rotor when disembarking from a helicopter: Bell 222.*

rotor is being stopped, you usually won't have to worry about this because the pilots won't start the shut-down procedure until they are certain all the passengers are clear of the aircraft.

I should mention one other very important caution. If you've landed in uneven or sloping terrain, be very aware of the rotor disc. The clearance from the disc to the ground will be less than normal on the upslope side of the helicopter. This might seem logical, but the clearance is even less than you'd expect because the pilot must tilt the rotor disc toward the slope in order to keep the helicopter planted firmly on the ground.

Watch your head. Don't become so excited about arrival that you don't pay attention when disembarking. And if you happen to be carrying any long objects, carry them horizontally, not vertically.

If the pilots shut everything down before you disembark, you won't have to worry about the rotor blades (FIG. 4-23). The noises, smells, and vibrations will be similar to

Fig. 4-23. *If the pilots shut down the helicopter before allowing the passengers to disembark, you don't have to worry about turning rotor blades, but you might hear some unusual noises.*

those experienced during start-up. Sometimes there will be a puff or two of smoke from the engines as they are shut off, but this is nothing to worry about as long as it doesn't last more than a few seconds. The whirring noise you might hear after everything has stopped is the spinning gyros inside the cockpit navigation equipment coming to a stop.

Your first flight is over. Do you still want to fly helicopters?

Good! Read on because the best is yet to come.

EMERGENCIES

Even though air travel is heralded as the safest form of transport, the frequent flyer, be he a pilot or passenger, is bound to experience some kind of emergency if he flies often enough.

Most aircraft emergencies are minor. Most delays or diversions due to mechanical problems are caused by the loss of one part of a redundant system. In other words, the aircraft can still fly safely, but one system, perhaps more, no longer have a backup. Aircraft design philosophy is to make all systems fail-safe. This is the philosophy that anything can fail at any time; therefore, everything should have a backup. If something fails, the flight can still be continued safely on the backup system.

Not everything in a helicopter is fail-safe. The main and tail rotors and the gearboxes do not have backups; therefore, these are subject to very frequent and thorough maintenance inspections. A large helicopter, for example, requires the same number of maintenance man-hours as a large passenger airliner four times its size.

In flight, the pilot has a number of warning systems that tell him if there is a problem with some part of the aircraft. Often, there is a ranking order to the warnings of a particular system. For example, an increase in oil temperature in a particular system might require no more than extra vigilance, but an increase in oil temperature with a decrease in pressure might require a landing within one hour; an increase in oil temperature with a decrease in pressure plus a warning light might mean an immediate landing.

Usually, the pilot has time to evaluate the emergency and decide on the best action. If he's doing his job and is not too busy, he'll tell the passengers what has happened, what he is doing about it, and what, if anything, they should do; however, in the event of a complete loss of engine power or loss of tail rotor control, the pilot will have to enter autorotation immediately. He will have his hands full and his first priority will be to make a safe landing; he probably won't have time to talk to the passengers and even a two-pilot crew might be so busy that they can't make a passenger briefing.

You can't mistake an autorotation for any other kind of descent. Normal descents are made at about 500 feet per minute. In autorotation, the descent is about 2,500 feet per minute, again depending upon many factors. On the way down, the pilot might have to bank the helicopter sharply a few times in order to head into the wind.

The best thing a passenger can do is prepare for a hard landing. Tighten your seat belt. Brace your legs on the floor. Memorize in your mind the location of the nearest emergency exit. Then lean forward and protect your head with your hands and your legs. With luck, the landing won't be too hard, but at least you'll be prepared if it is.

Helicopters have a big advantage over airplanes in that they can land vertically with little or no forward speed. They are therefore designed to absorb a great deal of energy from a vertical descent. The landing can be so hard that the structure of the helicopter is damaged or destroyed, but the passengers and crew will be able to walk away with no more than minor injuries.

Circumstances will dictate when you should disembark from the helicopter after an emergency landing. For example, if the pilot decides he must make a precautionary landing on a calm sea because of a suspected serious problem, the landing is safe and proper, and the helicopter is floating nicely, then there's no reason to leave immediately.

On the other hand, if the pilot must autorotate in mountainous terrain, the landing is very hard, and there is a lot of damage to the helicopter, it is probably better to disembark immediately. In such a case, your main concern is a post-crash fire and you'll want to get away from the aircraft as soon as possible.

One big caution when disembarking from a helicopter after a crash: Wait until the main rotor has stopped turning. It will only take seconds if the helicopter is on its side on the ground or in the water, but those seconds will make a big difference. In both cases the rotor disc will most likely be closer to the surface than normal. Wait until it stops.

Will a helicopter float if it lands on water?

Helicopters that fly often over long stretches of water are required by law to be seaworthy, like the amphibious S-61, be equipped with permanent floats (FIG. 4-24), or have flotation equipment installed (FIG. 4-25). These flotation bags can usually be inflated either manually by the pilots or automatically by an electrical mechanism.

But will the helicopter float?

Yes and no. Yes, if the sea is not too rough, it might float upright indefinitely. If the sea is too rough, it might float for a while, then turn upside down and float somewhat longer, or it might begin to sink fairly quickly. This is why life rafts for all people on board are also required for overwater flights.

Your best insurance is the pilots. I mean this seriously. They are not going to land on a rough sea unless it is absolutely clear that landing is the safest alternative, perhaps the only alternative.

A number of helicopters have ditched successfully in the North Sea, all on calm seas. Some people express amazement that all were so lucky that the seas weren't rougher; if you read the accident reports, it's evident that the pilots would not have ditched their aircraft if the sea had been too rough.

The cockpit indications were such that landing on a calm sea was a preferable alternative to continued flight; if the seas had been rough, thereby making a ditching much riskier, the preferable alternative would have been to continue the flight and take the risk of reaching land before the situation worsened.

Robinson Helicopter Company

Fig. 4-24. *Robinson R-22 equipped with floats.*

Fig. 4-25. *Emergency flotation equipment on Aerospatiale AS 332 Super Puma.*

What are the chances of surviving a serious helicopter emergency that results in an accident?

That's another difficult question to answer because it depends on so many factors. Every accident is different. All things considered, though, I believe your chances of surviving an accident in a helicopter are as good as, and perhaps better than your chances of surviving an accident in an airplane. Better because helicopters don't have to take off and land from long runways and statistics show that most aircraft accidents happen during these two phases of flight. And better because helicopter cabins are smaller and you'll probably be able to evacuate faster.

On the other hand, certain helicopter components, such as the rotor blades and gearboxes, are not fail-safe. A rotor blade separating from a helicopter is like a wing falling off an airplane. If that happens, the resulting accident is usually fatal. This is why these components on helicopters have numerous warning systems, are inspected thoroughly and frequently, and are replaced at very specific intervals. As a result, accidents due to failure of such items happen very, very rarely. If they happened more often, there would be fewer helicopter pilots around—as it is, there always seems to be plenty of applicants for the jobs that are available.

FINDING A RIDE

This might be the most difficult part of your introduction to helicopters.

If you wanted to experience fixed-wing flight as cheaply as possible, you could call up a travel agent and book a ticket on any one of a hundred or so commuter, charter, regional, or national airlines. With a little digging you could probably get a ticket for fewer than $50.

If you want to ride on a helicopter, your options are considerably fewer.

Fig. 4-26. *A few large cities have city-center to airport helicopter service: Agusta 109.*

Unfortunately, there are hardly any helicopter airlines around and the ones that are here today might be gone tomorrow. The history of helicopter airlines has been one dismal failure after another, with only a few notable exceptions.

If you live near a big city, you might find an operator who's plying the traffic between the city center and the airport (FIG. 4-25); however, your best bet might be a local sight-seeing flight (FIG. 4-26). Many big city operators have such flights. Operators in other areas of the country, for example near Niagara Falls and the Grand Canyon, do sight-seeing, too.

Fig. 4-27. *A flight with an operator that offers sight-seeing flights might be the best way to get an inexpensive ride in a helicopter: Bell 206B JetRanger.*

The best place to look for a helicopter operator is in the yellow pages of a telephone directory. If that doesn't yield results, try calling the local office of the FAA, typically referred to as a *flight standards district office*. You could try contacting a travel agent, but most agencies usually don't have much information about helicopters.

In the end, you might not find any helicopter operator in your area who sells tickets. Instead, you might have to charter a helicopter, which is essentially the same thing as if you were learning how to fly, except that you won't have to pay the cost of an instructor. The bigger the machine, the more it will cost to charter. Round up some friends that would like to go along and divide the charter expense.

Whatever you have to do to get a ride, do it. Believe me, it will be well worth your time and money.

5
Basic Flight Maneuvers

Nothing can take the place of persistence. Talent will not; nothing is more common than unsuccessful people with talent. Genius will not; unrecognized genius is almost a proverb. Education will not; the world is full of educated derelicts. Persistence and determination alone are the omnipotent.

Calvin Coolidge
30th President

WE'VE TALKED ABOUT LIFT AND ROTORS AND CONTROLS, YOU'VE GONE on your first flight, and now you have some idea how a helicopter works. Reading a book won't teach you how to ride a bicycle and it won't teach you how to fly a helicopter. But reading about any activity can give you a general idea how to do it and most people find this helpful. Because you probably already know how to ride a bicycle, I'll skip over that and go right into how to fly a helicopter.

Whatever your experience level, hovering should not be the first thing you try to do in a helicopter. That would be like trying to ride a unicycle on a tightrope before you've become good enough to take the training wheels off your bicycle. You may ask your instructor to demonstrate hovering flight during your first lesson so you can experience it, but your first chance at the controls of a helicopter should be in straight-and-level flight at a respectably safe altitude.

STRAIGHT-AND-LEVEL

First attempts won't be precisely straight-and-level flight, but it's a goal to work toward. Straight means you hold a constant heading; level means you hold a constant altitude (FIG. 5-1).

Your instructor should make the takeoff, climb to a safe altitude at least 1,000 feet above the ground, and turn the helicopter to a cardinal heading aligned with a prominent landmark near the horizon.

Fig. 5-1. *Straight-and-level flight should be the first maneuver you try in a helicopter: Schweizer 300.*

Cardinal headings north, east, south, and west are marked on the heading indicator N, E, S, and W, or 0, 9, 18, 27, for 0, 90, 180, and 270 degrees. People tend to get a little stressed when they're learning how to fly and round-numbered altitudes and cardinal headings are easier to remember and therefore easier to maintain when at the controls. By also using a prominent landmark on the horizon, you will be able to notice small heading deviations when you look outside the cockpit.

"I HAVE CONTROL"

When the instructor says, "You have control," position yourself comfortably in the seat, place your hands lightly but positively on the cyclic and collective, and rest your feet on the tail rotor pedals. Take a deep breath to calm yourself, and when ready, say "I have control."

Only when you say "I have control" will your instructor remove his hands and feet. Don't say it until you do have control. Actually, your instructor probably won't move his hands and feet too far away from the controls, because he knows his calming influence will be needed shortly, but within limits, you will be controlling the helicopter.

This "You have control," "I have control" procedure should not be trivialized. It is a basic, important control change procedure that you'll use throughout your career whenever you fly with another pilot, in any aircraft. It's the simplest way to ensure that at least one pilot always has his hands and feet on the controls and that both pilots know who is in control.

The pilot-in-command, or instructor in this case, will tell you when he wants you to take the controls by saying, "You have control." He'll keep his hands and feet on the controls until you say, "I have control."

When he wants to take the controls back, he'll say "I have control." You should check visually that he has taken the controls, say, "You have control," and then take your hands and feet away.

If you want him to take the controls while you're flying, say, "You have control," but don't release the controls, until he says, "I have control," and you've visually checked that he does.

If you would like to take the controls while the pilot-in-command or instructor is flying, be polite and ask, "May I have the controls?" or "Could I fly a while?" If the answer is yes, follow the above procedure. If the answer is no, don't force the issue. When you're second-in-command or a student, it's not your prerogative to take the controls unless told to do so or invited to do so by the pilot-in-command, except in an extreme emergency when it appears that the pilot-in-command isn't able to handle the situation and you believe you can do a better job.

If it seems that I am belaboring this point, it's only because I have learned from personal experience and the experience of others that confusion about who has control usually occurs at the most inopportune time and that it can be extremely embarrassing, not to mention dangerous, when both pilots think the other one has control and as a result, neither of them has it.

If your instructor isn't as fanatic about this as I am, do yourself a favor and use the procedure anyway. Your continued insistence on saying "I have control" and "You have control" might eventually embarrass him enough to start doing it, too.

The first thing that will probably happen after you say, "I have control," is nothing. If the aircraft is trimmed up properly for straight-and-level flight, it will pretty much stay that way and fly by itself without pilot input (FIG. 5-2). This will continue for several seconds, which will seem like minutes, and just when you think you're starting to get the hang of it, you'll notice you've started to turn off from the selected cardinal heading.

Fig. 5-2. *The helicopter will fly along quite well for a few moments without any input from the pilot: Bell 206L LongRanger.*

This is natural and to be expected because most of us (before we become helicopter pilots) are not too precise with our feet. What has happened is that in resting your feet on the pedals, you've inadvertently put a bit more pressure on one of them. This very correctly will cause the helicopter to yaw in that direction.

Your initial deviations probably won't occur with the cyclic or collective controls because you'll be concentrating much more on these. Actually, with a comfortable amount of friction on the collective and throttle, you can probably ignore your left hand. The cyclic will require the most attention and you'll be trying very hard not to change its position in the slightest from where your instructor left it. Eventually, no matter how much you initially try not to, you'll forget about your feet.

Next, you will try to correct for the change in heading. You'll probably do two things (if you've read this book, listened to your instructor, and are thinking at all sensibly). First, you'll push harder on one of the pedals and if lucky, it will be the one that turns you back toward the heading. Second, you'll try to help the machine back by adding cyclic in the direction you want to go. If the first action doesn't create problems, the second one will.

If you push the correct pedal, you'll probably push too hard. Well, what do you expect? You've never done this before. This initiates the turn back toward the heading too quickly and then you'll push too hard on the other pedal and then too hard on the first one again and back and forth until everything is totally out of whack and the instructor takes over. Believe me, by the time he says, "I have control," you'll want to give him the controls so quickly you'll forget to visually check that he really does have them before you release them.

If you make a cyclic input, you'll start rolling as well. And, because most people tend to pull back on the cyclic a little when they try to move it from one side or the other, you'll start some pitching movements, too. Pitching movements are when the nose starts to bob up and down. As it goes up—because you unintentionally pulled back on the cyclic—you'll try to counter this by moving the cyclic forward, probably too much. As the nose dips below the horizon, you'll counter again with back cyclic. And every time you try to correct pitch, you'll be adding more and more roll inputs to correct the first one you made to counteract the pedal movement you didn't mean to make in the first place.

PILOT INDUCED TURBULENCE

Everybody has experienced over-controlling or pilot induced oscillations and everybody does it. (Some pilots say "pit," for pilot induced turbulence.) Even experienced, professional pilots over-control when they check out in a new aircraft. It's almost impossible not to because every aircraft has a different feel.

The helicopter is going to be making all sorts of funny gyrations in the sky until you start to get a feel for it. Don't worry about it. It'll come. Ten hours of stick time is a fair rule of thumb before most pilots begin to feel comfortable in a new machine. But that's only a rule of thumb and everybody's different.

As a new pilot, with less experience to draw upon, it will take you longer, maybe three or four times as much, so don't worry if you don't catch on right away. Everything comes with practice. If you get discouraged, re-read the quote by President Coolidge at the beginning of this chapter.

So much for the pep talk, we're still trying to fly straight-and-level.

When the gyrations get too bad or the instructor feels your control movements are so much out of sync that you're not learning anything, he'll take the controls. In a second or so, the aircraft will be flying straight-and-level again, steady as a rock, and you'll swear your instructor is possessed with mystical powers. Don't let this bother you. It happens to everyone and your instructor just wants to give you a chance to start from a controlled position again.

One common instruction technique is to give the student only one or two controls to handle, while the instructor takes care of the other controls. This permits the student to concentrate on handling one control correctly while the other controls don't go to pieces. Because changes in one control can affect the others, it's easy to start to feel like a one-armed wallpaper hanger on a wobbly ladder until you get the feel for how much control input is needed for a given situation.

Watch what your instructor does to calm things down. It'll look like black magic at first, but basically what he'll do is simply put the cyclic, collective, and tail rotor pedals back to their neutral cruise flight positions and hold them there. When he wants to make a correction to come back on altitude or heading, he'll do it by increasing finger pressure on the appropriate control, not by moving it. Flying a helicopter takes an extremely fine touch. In fact, once you have trimmed up in straight-and-level flight, added a tad friction to the collective, and equalized the pressure on the pedals, you can easily fly by using only the thumb and forefinger of your right hand.

Finding the neutral positions and acquiring the necessary touch is what it's all about. It simply takes time and practice.

ACCELS/DECELS

A good way to improve proficiency in level flight is to perform accelerations and decelerations, or *accels/decels*. These will also help improve your "feel" of the aircraft and prepare you for other, more difficult, maneuvers, such as quick stops.

Accels/decels are very much like quick stops, although much gentler. The main differences between accels/decels and quick stops (*see* chapter 7) are that accels/decels are done with a slower rate of acceleration and deceleration, at altitude instead of close to the ground, and the lowest airspeed during the deceleration phase of the maneuver should be at or above the best-rate-of-climb airspeed.

Start out straight-and-level at the best-rate-of-climb speed for your helicopter, let's say 45 knots. Note the power setting you're using. You'll need it later. What you want to do is accelerate to high cruise speed without gaining or losing altitude or changing your heading. To gain airspeed, two things must be done: the nose must be lowered and the power must be increased. To lower the nose, ease the cyclic forward slightly. To increase power, raise the collective to maximum available power, adding throttle to maintain rotor rpm. Of course, as you increase power, the torque effect increases, too, and you must counteract this with left pedal.

As the airspeed increases, you'll have to make small adjustments to the cyclic position to keep from climbing or descending. When you reach high cruise speed, lower the collective to the high cruise power setting, reduce the throttle accordingly, adjust the cyclic to maintain level flight, and neutralize the pedals to remove any yaw tendency (FIG. 5-3).

Fly straight-and-level for a minute or two to settle yourself down. A deceleration is the opposite of an acceleration. To decelerate you must decrease power and raise the nose. To decrease power, lower the collective to the normal descent power setting, reducing throttle to maintain rotor rpm. To raise the nose, ease the cyclic back slightly. Of course, as you decrease power, the torque effect decreases, too, and you must counteract this by easing off left pedal pressure and maybe even adding right pedal.

Fig. 5-3. *Accels/decels are straight-and-level maneuvers that will improve your feel for the aircraft and prepare you for more difficult maneuvers later in flight training.*

As the airspeed decreases, you'll have to make small adjustments to the cyclic position to keep from climbing or descending. As you approach the best-rate-of-climb speed, raise the collective to the power setting you had before beginning the acceleration, increase the throttle accordingly, adjust the cyclic to maintain level flight, and neutralize the pedals to remove any yaw tendency.

The rate at which you increase and decrease the collective will determine how difficult the maneuver is. The faster you move the collective, the harder accels/decels are. Start off slowly. There's no reason to rush the maneuver the first times you do it. As you improve and feel more comfortable with the helicopter, gradually speed up the initial collective input.

Be careful not to allow the airspeed to fall below best-rate-of-climb speed, absolutely not below translational lift airspeed. There's no reason to tempt fate and an engine failure at a slow airspeed. Best-rate-of-climb airspeed is usually very close to the optimal autorotation airspeed; therefore, if the engine fails during the maneuver, it won't be difficult to enter autorotation at the best airspeed.

As soon as you start catching on to straight-and-level flight, your instructor will ask you to try some level turns. You might want to strangle him at this point, but don't try it because you need him to get safely on the ground. He is displaying some confidence in your learning abilities and believes it's time to step up to something more difficult.

If you steadfastly do not want to do any turns, say so. Instructors aren't gods or mind readers. The good ones could make pretty good psychologists and probably know better than you what you need at that moment, but other instructors might be plodding along adhering to a lesson plan regardless of your actual progress.

All instructors appreciate feedback. If you don't feel ready for a maneuver, request additional practice before moving on. Naturally, consider costs and do not waste time while simultaneously considering personal safety instincts. Consider requesting another demonstration of a maneuver to observe and integrate your experience to that point; that is an excellent chance for you to clear your head and relax hands and feet.

LEVEL TURNS

Recall that level means holding a constant altitude; turn means changing the heading. Turns come in all sizes, from one or two degrees through many times around the compass. Ninety-degree turns, from one cardinal heading to another, are a comfortable size to start off with. Students naturally progress from 90-degree turns to 180-, 270-, and 360-degree turns. Combinations for 540- or 720-degree turns are possible, but that's usually just boring holes in the sky. Obviously, turns can be made either to the right or left.

Before you do any maneuver, *clear the area*, look for any other aircraft above, below, and at your level in the direction of the maneuver. Make this a habit whenever you fly. Because of their slower speed relative to most airplanes, helicopters are easily overtaken and if your fixed-wing brethren aren't "seeing and avoiding" as well as they should, the encounter might be too close for comfort. Besides, helicopters provide much better visibility than most airplanes, so you probably have a better chance than the pilots in the airplanes of seeing that conflicting airplane traffic.

Before any turn training, clear an area left, right, and aft; do a pair of 90-degree clearing turns, first in the direction of the planned turn and second back to the original heading. You don't have to do clearing turns before every practice turn, but it's a good idea to do them every few minutes or so. Unless you happen to be training in a well-known practice area, pilots who do spot you might initially figure you'll be continuing in a particular direction, because this is the most common occurrence. They won't expect you to turn again and again, possibly toward their direction of flight, and might have, none too wisely, disregarded you. So make those clearing turns and watch out for yourself.

Making level turns in a helicopter is actually quite easy and very similar to making turns in airplanes. Like airplanes, you bank a helicopter toward the direction you want it to turn. Unlike an airplane, you don't need to use rudder to coordinate the turn (to prevent slipping and skidding) and counteract adverse yaw; however, because turns might cause changes in rotor rpm for aerodynamic reasons that are too complicated to get into here, you'll probably need to use some tail rotor pedal input to keep the turn coordinated.

Most texts on helicopter flying tell you not to use pedals in a turn. Theoretically the texts might be correct. In reality, I've found a little pressure on the left pedal when turning left and a little pressure on the right pedal when turning right tends to help keep the ball in the center. Don't ask for an explanation, it just works.

The most difficult part of level turns is staying level. When you tilt the rotor to one side to enter a bank, part of the vertical lift vector is lost. If you don't correct for this, you'll start to descend. To counter this loss of lift, you may either add power by increasing the collective or trade airspeed for altitude by easing back on the cyclic slightly.

If you enter a gentle bank and turn only a few degrees, you probably won't have to do either because the loss of lift will be minimal. If you enter a steep bank and do a complete 360-degree turn or more, you definitely will have to correct for this loss of lift.

You'll find that different helicopters act differently in turns. Some drop their nose considerably and require aft cyclic to keep them from entering a descent. Others tend to stay level in a turn, but lose airspeed fairly quickly. After flying one type for a while, you'll become quite used to its reactions and upon switching to another helicopter you might inadvertently apply an unnecessary correction.

TWO RULES OF THUMB

Two rules-of-thumb will serve you well when making turns:

First, angle of bank should not exceed one-half the number of degrees you wish to turn, up to a maximum of 30 degrees. In other words, if you want to turn 20 degrees, use a 10-degree angle of bank; if you want to turn 40 degrees, use a 20-degree angle of bank. If you want to turn 60 degrees or more, use a 30-degree angle of bank.

Second, start the roll-out to the desired heading using one-half the number of degrees in the bank angle. For example, you're turning right from 090 degrees to 270 degrees using a bank angle of 30 degrees. Start the roll-out at 255 degrees, 15 degrees prior to reaching 270 degrees. If done smoothly and steadily, this should put you within a degree or two of the desired heading every time. When that close, you may get rid of that last degree by using the pedals, but do it gently to avoid noticeable yaw.

Just when you think you're getting the hang of level turns, your instructor demonstrates climbs and descents.

NORMAL CLIMBS

A helicopter climbs three ways: pull back on the cyclic; raise the collective; and pull back on the cyclic and raise the collective.

Pulling back on the cyclic is similar to pulling back on the stick or yoke in an airplane. The nose goes up, airspeed goes down, and the aircraft starts to climb. You're trading airspeed for altitude. If you pull the nose up too much, the helicopter will climb very quickly initially, but, just as quickly, airspeed will decrease; in an airplane you risk entering a stall; in a helicopter you can bleed airspeed all the way to zero and perhaps end up, depending upon aircraft weight and the power setting, either moving backwards in a high out-of-ground effect hover or in a nose-up shallow descent.

Pulling up the collective of a helicopter is akin to adding power in an airplane, but differences are apparent; in an airplane, the airspeed increases; in a helicopter, you start to climb and airspeed stays the same. (If you wanted to increase airspeed and not climb in a helicopter, you must also add forward cyclic when you pull up the collective.)

Finally, the most effective way to climb is pulling back on the cyclic and increasing the collective.

The third method is the commonly accepted way to climb, but the other two ways are acceptable under some circumstances. Let's say, for the sake of an example, you're a military pilot flying nap-of-the-earth at maximum airspeed and with full power (FIG. 5-4). You're approaching a line of trees that you'd rather fly over than around. You can't increase the collective pitch any more because the collective is already in your armpit. So, you ease back on the cyclic, zoom up to just above tree-top level (losing maybe five knots in the process), and ease the cyclic forward again to level off. Within seconds, you regain the five knots you lost in the climb and everything is the same as before, except you're a few feet higher over the ground.

Fig. 5-4. *By doing a cyclic climb, the pilot causes the helicopter to climb quickly by trading airspeed for altitude: Sikorsky MH-60 Pave Hawk.*

CYCLIC-ONLY CLIMBS

Such altitude-for-airspeed climbs, often called *cyclic climbs*, are good for gaining small amounts of altitude quickly: actually, the fastest way to gain altitude in a helicopter. The disadvantage is that the initial high rate of climb will dissipate rapidly as airspeed drops off. It all depends upon the airspeed when the climb is initiated and how much airspeed can be lost in the climb. To achieve the best climb, allow the airspeed to drop to best-rate-of-climb speed and hold it there. If any slower, climb rate is reduced. For the sake of safety, don't let the airspeed drop below translational lift airspeed.

One important thing to remember about cyclic climbs: Do not pull back on the cyclic too quickly. The faster you pull back, the more g-forces are applied. Civilian helicopters are not built to take as much g-force as most airplanes and selected military helicopters, and something might break.

Of equal concern is the flapping tendency of the main rotor blades. When you pull back on the cyclic, the back of the rotor disc tips down. With a high positive g-force also acting on the rotor disc, there's a good possibility that one or more main rotor blades might contact the tailboom or even sever the tail rotor drive shaft. So, when you do a cyclic climb, do it with care and don't try to make a vertical climb with the little rotary-wing machine. It's not a space shuttle.

COLLECTIVE-ONLY CLIMBS

Climb method number two, the collective-only climb, is useful if you want to maintain airspeed, don't need to climb quickly, and don't have to gain more than 500 feet or so. As such, it's a good method for climbing when you are flying on instruments, for example, and air traffic control requests an altitude change (FIG. 5-5). It's also useful to maintain altitude in minor turbulence.

Fig. 5-5. *Collective-only climbs are a good way to climb when flying IFR or when you don't need to gain much altitude: Agusta A109MAX.*

Agusta Aerospace Corporation

You'll normally cruise at a power setting equivalent to approximately 60 or 70 percent of full power; thus, you'll have ample power available to accommodate a climb of at least 500 feet per minute while maintaining cruising airspeed. Collective-only climbs are probably the most comfortable for passengers because fuselage attitude does not vary. They probably won't even notice the altitude change.

One disadvantage of the collective-only climb is that the rate of climb usually is not that great, and might be downright marginal if the aircraft is at maximum gross weight and it's a hot day. Also, because you're maintaining airspeed and climbing, you might have to use full power, which is not ideal for the engine if you have to gain a lot of altitude. And more fuel is consumed.

BEST CLIMB METHOD

The third way to climb is the most popular because a cyclic and collective climb is the quickest and most fuel efficient, advantageously combining the other two methods with none of the disadvantages. You can initiate the climb quickly by easing back on the cyclic, trading airspeed for altitude, plus sustain the climb and maintain a comfortable airspeed by increasing collective pitch. Because the climb is at a lower airspeed and the rate of climb is greater than with a collective-only climb, you don't have to maintain the climb as long and the wear-and-tear on the engine is minimized.

Assume you're cruising along at 100 knots straight and level and at a fixed power setting. To start the climb, you ease back on the cyclic to bring the nose up five to 10 degrees above the horizon. You could start the climb by increasing the collective first and then easing back on the cyclic, but this won't initiate the climb as quickly as making the cyclic input first.

You should see an immediate indication on the vertical speed indicator showing a climb and the altimeter will slowly increase. Airspeed reacts, too, by decreasing. If you hold the nose up and don't do anything with the collective, airspeed will continue to bleed off until it stabilizes at a slower speed that corresponds to that particular nose-up attitude and that power setting. Depending on conditions and gross weight, you could end up either level or descending. Instead of waiting to let that happen, you should increase the collective pitch to the helicopter's climb power setting.

Now you must adjust the other controls to keep everything working together properly. An increase in collective means blade pitch increases; therefore, in order to maintain rotor rpm, engine power must also be increased, a *power on*, by rotating the throttle counter-clockwise (as viewed from the pilot's seat). As collective pitch and throttle are increased, so is the torque effect of the main rotor and the nose of the helicopter yaws to the right. Increased pressure on the left pedal is needed to keep the machine on heading.

FLYING WITH YOUR EARS

One thing we haven't talked about up to now is sound. Your ability to hear changes in rotor rpm and engine noise is going to become very important to you as a helicopter pilot. You won't notice it so much at first, but as you gain time in a machine, you'll learn what normal rotor rpm and engine noise sound like. You'll also be able to distinguish between high and low rotor rpm and unusual engine sounds. Many experienced helicopter pilots will hear critical changes in rotor rpm and engine rpm before noticing the changes on the gauges, a very effective early warning system.

Don't try to learn these sound indications on your first few flights—you'll have enough else to do—but do be aware of them. I mention them now because during climbs and descents, when you are necessarily making changes in the collective and throttle settings, you'll be exposed to differing sound indications from the rotor and engine.

The airspeed you use in the climb will depend on the helicopter type and how quickly you want to climb. Two climb airspeeds are often stipulated by the manufacturer. The best-rate-of-climb airspeed (V_y) is what it says, the airspeed that yields the best (greatest) rate of climb. Learn and do not forget this airspeed because it offers the absolute best climb performance for that helicopter. Actually, V_y varies plus or minus a few knots depending upon atmospheric conditions, but the change is so slight that you can safely use one speed. Besides, airspeed indicators aren't that precise anyway.

The other airspeed often, but not always, given to us by the manufacturer is the cruise climb airspeed. This is always higher than V_y and doesn't give as much climb—only one speed can be the best—but it will give you an acceptable climb without requiring you to slow down all the way to V_y. In helicopters with relatively high cruise speeds, the V_y might be as low as 50 percent of the cruise climb speed.

If the manufacturer hasn't established a cruise climb speed, pilots and operators often pick a speed that gives them good climb performance without sacrificing too much airspeed.

Anticipate leveling off to end up at the desired altitude without busting through. Use the 10 percent rule of thumb. If climbing at 500 feet per minute, start to level off when 50 feet below the desired altitude. If climbing 1,000 feet per minute, start to level off 100 feet below the desired altitude. If climbing 5,000 feet per minute, you're not flying a helicopter.

Ease the cyclic forward first. This will cause airspeed to increase and the rate of climb to decrease. Then, as the airspeed approaches cruising speed and the altimeter nudges toward the desired altitude, reduce the collective to the cruise power setting. Reducing the collective will necessitate a reduction in engine power in order to keep rotor rpm from increasing. Rotate the throttle clockwise as you lower the collective. Torque is consequently reduced; therefore, keep the nose straight, releasing the extra pressure you put on the left pedal to counteract the increase in torque during the climb.

If you've climbed no more than 1,000 feet or so, power setting, airspeed, control positions, and overall vibration level will be approximately the same as the previous altitude; if you've climbed 2,000 feet or more, you'll definitely notice a difference in the above parameters. With the same power setting, indicated airspeed will be less (although true airspeed might be the same or higher depending on atmospheric conditions). To get the same indicated airspeed, you'll have to use a higher power setting. The collective lever will be raised, the cyclic will be more forward, you'll need more pressure on the left pedal, and the overall level of vibrations will be greater.

As you gain more experience and confidence in the helicopter, take the time to note outside air temperature before and after the climb. Usually, you'll see it decrease

2 °C per 1,000 feet, which is the *standard lapse rate*. Sometimes temperature will decrease quicker, indicating the presence of colder-than-normal air above, possibly due to an approaching cold front. An increase in outside air temperature while climbing indicates a temperature inversion. Inversions are often indicative of future fog or smog, if not already present.

More than likely, after climbing a few thousand feet, your instructor will request a descent.

NORMAL DESCENTS

Straight ahead normal descents are probably the easiest maneuver to do in a helicopter (FIG. 5-6). Perhaps only the pull of gravity makes descents seem this way. Perhaps it's only in the mind of the pilot, a human factors psychological issue that somehow relates descents in aircraft to coasting down hills on a bicycle on a warm summer day. Perhaps it's because they're just easy to do.

Fig. 5-6. *Normal descents in a helicopter are probably the easiest maneuver to do: Sikorsky S-76B.*

There are two basic ways to initiate a descent: lower the collective or apply forward cyclic.

The first way is more common and will allow a much wider range of descent rates than the second way. You can vary the rate of descent from as little as a few feet per minute to full autorotational rate of descent (about 2,000 feet per minute or more, depending on conditions) just by using the collective.

On the other hand, using forward cyclic to start a descent is limited by the never-exceed airspeed (V_{ne}). If you're cruising only 10 or 15 knots below V_{ne} when you start the descent with forward cyclic, you'll soon be bumping the limit.

To start a normal descent, simply lower the collective an inch or two. Your instructor should recommend an approximate power setting to use. Of course, every time you move one control in a helicopter, you must adjust one or more of the others. In this case you'll have to rotate the throttle clockwise, a *power off*, to keep the engine from revving the rotor rpm too high and apply pressure to the right tail rotor pedal to keep the nose straight. You need right pedal because the nose will yaw to the left because of the reduction in torque when you lowered the collective.

You might have to make an adjustment to the cyclic position, although theoretically this shouldn't be necessary. Airspeed will tend to increase a few knots, but this shouldn't be any problem unless you happen to be cruising at close to V_{ne} speed, which is unlikely. If airspeed does build too much, simply ease back on the cyclic a tad and that should stabilize it. If this aft cyclic movement slows the rate of descent too much, simply lower the collective a bit more.

Use the same 10 percent rule of thumb as a guide to determine when to start leveling out. If descending at 700 feet per minute, start to level out 70 feet above the desired altitude.

Level out by increasing collective back to cruise power, add throttle to maintain rotor rpm, and ease the pressure off the right pedal. If you made an aft cyclic correction to keep the airspeed in limits on the way down, slowly ease the stick forward again to maintain cruise airspeed.

That's really all there is to it.

TURNING CLIMBS AND TURNING DESCENTS

Not to worry you, but these maneuvers will probably seem very difficult the first times you do them. It's only because you'll have to do a number of things simultaneously and that always takes some getting used to. But learning climbing turns and descents is well worth the time and effort because they are common maneuvers, particularly in the traffic pattern.

Learn how to perform numerous tasks simultaneously by doing the individual tasks in sequence by the numbers. As you become more and more proficient at each individual task, repeat the actions faster and faster and eventually you'll be able to do all the tasks at the same time.

That's why you start off straight-and-level, progress to level turns, continue to climbs and descents, and finally end up with climbing and descending turns, building skills along the way to acquire proficiency in the individual tasks before trying to put them all together.

A climbing turn is really nothing more than a level turn with a climb thrown in, or a climb with a level turn thrown in. You don't do anything more than you've learned up to this point; you're just combining two maneuvers into one. In theory, it shouldn't be all that difficult, but it will seem hard because you won't be accustomed to juggling so many elements at once.

I suggest you do it by the numbers until you get a good feel for it. Figure 5-7 gives step-by-step procedures for climbing and descending right and left turns.

	RIGHT CLIMBING TURN	LEFT CLIMBING TURN	RIGHT DESCENDING TURN	LEFT DESCENDING TURN
1	Clear the area for other aircraft			
2	Visualize the maneuver			
3	Right cyclic	Left cyclic	Lower collective	Lower collective
4	Increase collective	Increase collective	Decrease throttle	Decrease throttle
5	Increase throttle	Increase throttle	Right cyclic	Left cyclic
6	Some right pedal	Definite left pedal	Definite right pedal	Some left pedal
7	Check the gauges: Airspeed, vertical speed, heading, altitude, power			
8	Adjust cyclic, collective, throttle, and pedals as required			
9	Repeat steps 7 and 8 throughout the maneuver			
10	Begin to level off and roll out using rules of thumb			
11	Center cyclic	Center cyclic	Raise collective	Raise collective
12	Decrease collective	Decrease collective	Increase throttle	Increase throttle
13	Decrease throttle	Decrease throttle	Center cyclic	Center cyclic
14	Ease off right pedal	Ease off left pedal	Ease off right pedal	Ease off left pedal
15	Check the gauges: Airspeed, vertical speed, heading, altitude, power			
16	Adjust cyclic, collective, throttle, and pedals as required			

Fig. 5-7. *Left and right climbing turns by the numbers.*

DOING IT BY THE NUMBERS

Clear the area to check for other traffic, then form a mental picture of the next maneuver, for example, a 180-degree left climbing turn. Okay, visualize the helicopter turning left and climbing to your rear (FIG. 5-8). What steps do you have to do? (FIG. 5-7):

3. Start the turn with left cyclic.

4. Increase the collective to climb power.

5. Increase the throttle to maintain rotor rpm.

6. Keep the turn coordinated by adding left tail rotor pedal.

7. Check the gauges: airspeed, vertical speed, heading, altimeter, power.

8. Adjust the cyclic, collective, throttle, and pedals as needed to maintain a constant rate of climb and turn.

9. Repeat Steps 7 and 8 throughout the maneuver until it's time to level off and roll out.

10. Begin to roll out using the rules of thumb for turns and climbs.

11. Return the cyclic to the center or neutral position.

12. Decrease the collective to the cruise flight position.

13. Decrease the throttle to keep rpm in limits.

14. Ease off the added pressure on the left pedal.

15. Check the gauges to ensure you've ended up where you want to end up.

16. Make a final adjustment to the cyclic, collective, throttle and pedals, if needed.

It sounds complicated when you read it, but, believe me, do it by the numbers enough times and you'll soon be doing it instinctively, smoothly, and simultaneously. It will still take concentration and attention to what you're doing, but it won't seem nearly as difficult.

Use the similar by the numbers procedures as outlined in FIG. 5-7 for right climbing turns, right descending turns, and left descending turns.

Notice that climbing turns are started with cyclic input whereas descending turns are started with collective input. This is a matter of technique based on personal preference and experience in the helicopter types I have flown. It's not cast in stone. I feel comfortable initiating the maneuvers this way. Your goal should be to do steps 3 through 7 and steps 11 through 15 more or less simultaneously. If you feel more comfortable starting climbs with the collective input and descents with the cyclic input, feel free to do so.

It will probably only take you a few hours to master straight-and-level flight, level turns, and normal climbs and descents. By the time you've done these, you'll start to have a feel for the machine. Feel is hard to define, but you'll know it when you get it. It's akin to learning how to balance on a bicycle.

McDonnell Douglas Helicopter Company

Fig. 5-8. *AH-64 Apache making a left climbing turn during Operation Desert Storm. Do climbing and descending turns by the numbers until you can do them instinctively.*

Climbing turns and descents will take many hours of practice before you can do them well, but as mentioned before, it's worth the time and effort. Fortunately, you don't have to be completely proficient in climbing turns and descents before progressing on to the next phase of your training, hovering, which is perhaps the most exciting part of flying a helicopter.

6
Learning to Hover

I did most of the flying at that time and became very familiar with the helicopter's operation. During my years in aviation, I had never been in a machine that was as pleasant to fly as this light helicopter was, with a completely open cockpit. It was like a dream to feel the machine lift you gently up in the air, float smoothly over one spot for indefinite periods, move up and down under good control, and move not only forward or backward but in any direction. As for landings, it was possible to come down not only within a few feet but even within a few inches of a spot previously designated on the ground and this was easily done, even with rather strong winds.

Igor Sikorsky,
describing the VS 300 in 1940;
his book, *The Winged S*

HOVERING MAKES A HELICOPTER A HELICOPTER. THAT MIGHT SOUND corny, but it really is true. If a helicopter couldn't hover, it might as well be an airplane. Or to put it another way, if your job doesn't require an aircraft that can hover or an aircraft that can take off from and land in a very small space (which essentially requires hover capability), you don't need a helicopter.

Hovering is the "raison d'être" of helicopters. It's the main advantage helicopters have over airplanes. It's the single most important rotary-wing capability that keeps helicopter operators all over the world in business (FIG. 6-1).

Helicopter Service Bell 212

Fig. 6-1. *Picking people off a fishing boat with a hoist is a job that can only be done by helicopter.*

It's also fun.

Cars in parking lots at helicopter flight schools abound with bumper stickers that say, "Hover Lover" and "To Fly is Human, To Hover is Divine." You can't help getting a kick out of hovering.

You sit there motionless, a few feet over the ground, and unattached to anything earthbound. Want to see what's behind you? Press on a pedal and you turn around. What's that over there? Nudge the cyclic and you slide over to it. Can't see what's on the other side of that fence? Lift the collective and you climb like an elevator. No king ever sat on a more wonderful throne. It's great.

It's also hard.

It's the hardest thing you'll have to learn. But once you master it, you'll have it. Like riding a bicycle.

In fact, a pilot friend of mine likes to use a bicycle analogy when describing helicopter flying. He says flying an airplane is like riding a bicycle and flying a helicopter is like riding a unicycle. It's not a bad analogy.

There are a lot of similarities between airplane flying and helicopter flying, just like riding a bicycle is similar to riding a unicycle. But, riding a unicycle requires something more, too—more skill, better balance, greater concentration, more practice. Flying helicopters, and particularly hovering, is like that.

If you've flown airplanes, hovering will be a new, slightly disconcerting sensation. If you've never flown before, it will simply be awesome.

THE BASIC HOVER

It's important to remember that a hovering helicopter is a flying aircraft, even though it is stationary over one spot. The helicopter might only be a few inches above the ground, but it is now a creature of the air and all aerodynamic rules and principles apply. I say this because a hovering helicopter might look stable and easily controlled; it might be stable, but it's not necessarily easily controlled. The pilot is working hard, sometimes very hard, to keep it where it is.

Before lifting into a hover, check that the cyclic and tail rotor pedals are in their neutral positions, in other words, that you haven't inadvertently pushed the right pedal forward slightly or moved the cyclic stick one way or the other. On the ground, moving these controls will generally not move the helicopter, but anytime the rotors are turning you must pay attention to any control inputs. By checking the position of these controls immediately before takeoff, you help ensure that the helicopter will lift straight up and not veer or turn to the right or left.

Clear all around, right, left and above. Remember, you're going up and you don't want to hit someone flying low over the top of you. You can never be sure the way is clear, so check overhead. Check the sides for people, vehicles, and other aircraft (FIG. 6-2).

When you're ready to go, lift the collective slowly upward while at the same time increasing engine power with the throttle, keeping rotor rpm within limits. As the

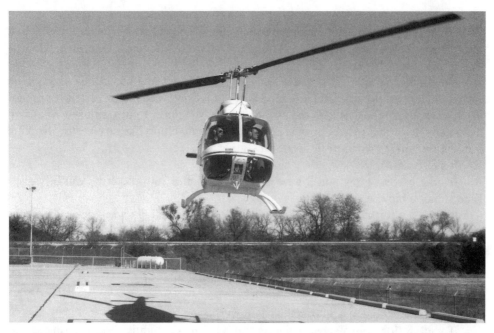

Fig. 6-2. *A Bell 206 pilot clears the area, checking for traffic, prior to a practice maneuver.*

pitch on the main rotor blades increases, the lift increases and the helicopter becomes *light on the skids* or wheels. The machine is now half-flying and half on the ground and you'll have to make careful adjustments with the cyclic and tail rotor pedals. In American-made helicopters, whenever you raise the collective, the fuselage will want to turn to the right, so you can expect to progressively increase the pressure on the left pedal as you lift into a hover.

This transitional phase when the helicopter is not quite flying and not quite on the ground requires extreme vigilance. Two nasty things can happen if the pilot is not careful: *ground resonance* and *dynamic rollover*.

Ground resonance occurs when the pilot's collective inputs get out of synchronization with the aircraft and the machine starts bouncing up and down. The springiness of the skids or landing gear only aggravate the situation. Ground resonance usually occurs when the pilot is trying to be too precise and too cautious and starts pumping the collective up and down. A good way to avoid pumping the collective is to increase the friction on the collective pitch lever by rotating the friction lock in the proper direction. This will make the collective seem heavier and harder to move.

Dynamic roll-over is mainly a problem when taking off from a sloping surface or with a crosswind. It is caused by too much lateral cyclic, which is an easy mistake when you're trying to hold the machine steady on a slope. The problem is you could cause the machine to enter a condition in which it begins to roll over while balanced on one skid and go past the point where no amount of opposite lateral cyclic will counteract the roll.

The way to avoid both ground resonance and dynamic roll-over is to be sure the cyclic is in the neutral position when light on the skids and to pull the collective up in one smooth motion. It shouldn't be a fast, jerking movement, but rather a steady, constant-rate pull. Whatever you do, don't stop halfway between firmly on the ground and fully in the air.

As you pull the collective upward, one skid or main wheel will leave the ground first, not because the helicopter has been loaded improperly but because of the way it is constructed. To improve forward flight performance, the main rotor mast is tilted a few degrees forward. The tilt and the gyroscopic effect cause the helicopter to hover with a slight bank in a no-wind condition, a compromise that is acceptable. It should also be noted that a crosswind can exaggerate the bank, eliminate it, or even cause the helicopter to bank in the opposite direction.

The tendency of most helicopters to lift into a hover with one gear lower than the other is another reason you don't want to be too prim with your upward collective movement. If you dally too long with one wheel or skid touching the ground, sooner or later the helicopter is going to want to move. The wheel or skid will act as a pivot point and cause the machine to pirouette, but not very gracefully.

The hardest part of hovering will be pilot-induced turbulence (PIT), your own erratic and unnecessary control inputs. Try to calm down. That's easy to say, but hard to do. You will be clutching the controls and your instructor will tell you to relax your grip. That's easy to say and hard to do, too. Believe it or not, you'll eventually be able to hover using only finger pressure.

Although the cyclic will feel about as firm as a wet noodle and you'll be concentrating a lot on it, the biggest offender is often the collective. It is the one control that will always create the need for correcting inputs from the other controls, so calm it down first. Try to find a fixed power setting that will hold a comfortable hover and leave the collective at that setting; at most, make very small adjustments. Whatever you do, don't pump the collective up and down (FIG. 6-3).

Stabilizing the collective will eliminate a multitude of other sins. Now you can adjust the throttle and leave it set. After that you can pay less attention to the pedals because you won't be changing torque. Then you can concentrate on that wet noodle.

It's very important to find a comfortable seat height and position so that you can rest your right forearm on your right thigh. This will give you a stable platform from which you can control the cyclic. Cyclic movements should be done from the wrist, not the elbow. Actually, the movements needed are so precise, that they are more on the magnitude of pressures than movements. If you find yourself working the cyclic like an old-fashioned butter churn, your movements are much too big. All you're doing is creating PIT and working against yourself.

Your first attempts at hovering will be worse than your first attempts at straight-and-level flight. I think you will appreciate the wisdom of getting a feel for the

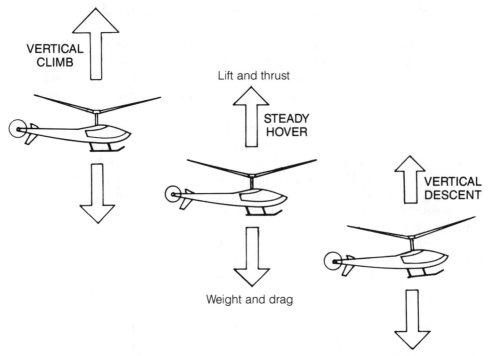

Fig. 6-3. *Forces acting upon a helicopter in a hover, no wind condition. Increasing the collective increases lift and causes the helicopter to ascend vertically. Decreasing the collective decreases lift and causes the helicopter to descend vertically. Holding the collective constantly, stabilizes the hover at a precise height. (Blade coning angle exaggerated in drawing.)*

machine in cruise flight before attempting to hover. Hovering takes practice, lots of practice, so don't be discouraged.

If you've flown airplanes before, an excerpt from an article by J. Mac McClellan in FLYING Magazine, might raise your spirits:

"I believe the experienced airplane pilot has a small edge (over the nonpilot) when transitioning to helicopters, though learning to fly helicopters is the most challenging and difficult aviation task I have ever faced."

A FEW TRICKS OF THE TRADE

Helicopter pilots use a few tricks of the trade to hold a steady hover over one spot. One is to use two or three hover references instead of concentrating on only one spot. Pick one hover reference about 20 to 30 feet in front of you, another at a 45-degree angle to the side at about the same distance, and a third between the two a few feet away. Move your eyes from one reference spot to another and occasionally bring the horizon into your field of vision, too. If you look only at one point, it will be difficult to see small changes in attitude.

Try to think of the entire windshield as a big attitude indicator. Heading control will be easy because small deviations will be readily apparent on the horizon. Peripheral vision will give you depth perception and help you detect movement to the sides without having to turn your head.

One of the best tricks involves the "gun sight technique." First, pick an object relatively close to your position, for example a small tree or pole, and then line one point of that object onto another object in the distance. The trick is to hold the first object on top of, or in the same relation to, the second object. Using one "gun sight" works fairly well, but you can still end up moving forward or backward along the line of the sight; therefore, it helps to have another gun sight at an angle to the first one. Very small movements of the helicopter are very easy to detect using this method and by correcting the small deviations quickly, you avoid the big ones.

HOVERING TURNS

Hovering turns are relatively easy in no-wind conditions, once you've mastered the ability to hover over one spot. In theory, all you need to do is to add pressure to the pedal on the side you want to turn toward. Push the right pedal forward and you move to the right; push the left pedal forward and you move to the left.

Of course, as we have seen before, making an input on one control usually necessitates a correcting input on one or more of the other controls. Hovering turns are no different.

In a stable, no-wind hover everything is in equilibrium. The engine is providing just enough power to keep the main rotor turning at just the right rpm to hold the height over the ground and to keep the tail rotor spinning at just the right angle to counteract the torque of the main rotor. When you push one or the other of the tail rotor pedals forward, you upset this balance and must do something to compensate in order to hold the same hover height.

In a hover, the torque of the main rotor tries to turn the fuselage to the right and therefore pressure on the left pedal is required to keep the nose straight. Forward left pedal means the tail rotor blades are biting the air at a greater angle of attack and therefore producing more lift. If you add even more forward left pedal to initiate a left hovering turn, you increase the tail rotor blade pitch angle even more. This requires more power from the main gearbox and, if engine power remains constant, the only way the gearbox can satisfy this increased demand for power from the tail rotor is to allow main rotor rpm to decrease.

If you push left pedal while hovering and do not compensate with collective and throttle, the helicopter will not only begin to turn left, but will also begin to descend as rotor rpm decreases. You might not descend all the way to the ground if you are very gentle with the left pedal, but you will probably get close.

The correct procedure, then, is to counter the increased demand on the main gearbox by increasing the throttle to maintain rotor rpm when turning to the left.

A right hovering turn causes just the opposite to occur. When you push right pedal, the tail rotor blade pitch angle decreases and demands less power from the main gearbox. The result is that more power is made available to drive the main rotor blades and rotor rpm will increase. When rotor rpm increases, the helicopter will climb vertically. It usually won't climb very much, but if you're trying to hold a precise hover height, just a few feet higher might be too much.

The correct procedure, when making a right hovering turn, is to decrease the throttle slightly in order to keep rotor rpm from building.

Throughout the maneuver you should maintain position over the ground with the cyclic. You probably won't need to adjust the cyclic very much with no wind, provided you make pedal and throttle movements smoothly and correctly. PIT sends the helicopter gyrating all over the sky.

Another thing you should know about hovering turns is that the helicopter will rotate around the main rotor mast when you make a pedal-only hovering turn. In a small helicopter, the pilot sits with his back almost against the mast and it's easy to get the impression that you are at the center of rotation (FIG. 6-4). For all practical purposes you are, but if the pilot's seat could scribe its position on the ground as you turn, you'd find it made a small circle as opposed to a point.

Fig. 6-4. *In small helicopters, like this Schweizer 300, the rotor mast, and therefore, the axis of rotation in a hovering turn, is directly behind the pilot's back.*

In larger helicopters, the pilot sits farther away from the main rotor mast and the pilot's seat would scribe a much larger circle when doing a 360-degree hovering turn. The difference is like sitting on the edge of a phonograph record or near the spindle. As a consequence, two hovering turns are often used in larger helicopters.

One was just defined, simple rotation around the main rotor mast. The second is more difficult because it always requires cyclic inputs, a good deal of them. This is the hovering turn in which the axis of rotation is not the main rotor mast, but some other point. The point could be the pilot's seat, or the nose of the helicopter, or even an imaginary point in front of the nose. The main rotor mast would scribe a circle over the ground. You can imagine the complexity of control inputs required for such a maneuver.

As you gain proficiency, hovering turns around an imaginary point in front of the nose are a good training maneuver to improve overall control coordination.

HOVERING WITH WIND

Up to now, we've only talked about no-wind conditions. In most parts of the world, some wind is usually present, so we can't expect to encounter no-wind conditions very often; however, wind does have its good sides, too.

Hovering a helicopter requires a lot of power; normally a helicopter will use less power at cruise speed in forward flight than it does hovering; at the best endurance airspeed a helicopter might use as little as 50 percent of the power that it uses to hover.

The reason is translational lift, discussed in chapter 4. Translational lift, as you remember, is the additional lift obtained by the helicopter because of increased efficiency of the rotor system as the helicopter moves from hovering into forward flight or when it hovers in a wind. The rotor disc, which is the circular area defined by the sweep of the rotor blades, behaves as if it were a solid wing and picks up extra lift that supplements lift created by the rotation of the blades.

Wind, then, improves the helicopter's ability to hover by simulating, if you will, forward flight. This assumes, of course, that the helicopter is hovered with the nose into the wind, which is usually the case. Hovering in a sidewind or tailwind requires more power than a no-wind hover and is therefore only done if there is an operational necessity.

Hovering with the nose pointed into the wind is actually easier than hovering with no wind. The weather vaning tendency of the helicopter and the added lift from the wind help you hold a more stable hover.

On the other hand, when you want to make a hovering turn, things become more difficult. In fact, helicopters have crosswind and tailwind hovering limitations. If the wind exceeds these values, hovering turns should not be done.

As soon as you turn the nose, the wind will start to push the helicopter downwind. The more you turn, the more it will push. To maintain position over the ground, you must tilt the rotor disc into the wind, which requires cyclic pressure toward the wind. As you turn, you must continually readjust the cyclic position.

Hovering over one spot with a 10-knot headwind is the same as moving forward at 10 knots in a no-wind condition. As you turn 90 degrees, it will be the same as hovering sideways at 10 knots. Turn the tail into a 10-knot wind, and it's the same as hovering backward at 10 knots in a no-wind condition. So you can see that when you do a 360-degree hovering turn when it's windy, you're not only turning, you're also transitioning from forward to sideways, to rearward, to sideways, and back to forward hovering flight.

HOVERING FORWARD, SIDEWAYS, AND REARWARD

Hovering over one spot is fine and being able to turn 360 degrees is quite useful, but helicopters wouldn't be worth much if they couldn't move from one place to another while still hovering.

Hovering forward is the easiest, perhaps because it is the most natural. You're facing forward, you're probably used to driving a car (mostly) forward, and if you've flown airplanes, you've done that going forward, too.

From a stable hover, ease the cyclic forward very slightly. This will tip the rotor disc forward, cause the nose to dip slightly, and initiate movement. It will also cause you to descend a bit because now you're using part of the total lift vector as forward thrust, whereas before it was only being used to counter the weight of the machine. To avoid descending, increase collective a small amount and simultaneously increase throttle to maintain rotor rpm. Of course, when you increase collective and throttle, you also increase torque, so you'll need left pedal pressure to counter the fuselage's tendency to yaw right.

As soon as you start to move forward, ease the cyclic back to it's stable hover position. If you don't, you'll keep accelerating and eventually pass through translational lift and enter forward flight.

To stop hovering forward requires an aft (rearward) cyclic application. This tips the rotor disc and the nose upward and slows the machine. As you come to a stop, remember to reduce the collective to its original hover power setting and adjust throttle and the pedals as necessary to maintain rotor rpm in limits and stay on heading, respectively.

Like I said, hovering forward is the easy one.

Hovering rearward is the opposite of hovering forward, almost. The required cyclic control inputs mirror what you do when hovering forward. Instead of using forward cyclic and lowering the nose, you use aft cyclic and raise the nose. The collective, throttle, and pedal inputs are the same: up collective to counteract the loss of some of the vertical lift vector to a horizontal vector, increased throttle to maintain rotor rpm, and left pedal to counteract the increased torque. The thing that makes it different is the position of the tail rotor.

With the tail rotor now leading the rest of the helicopter, you have an unstable longitudinal situation, not unlike wind blowing on a weather vane. Quite simply, the tail feels uncomfortable out front and very much wants to take its rightful place in the rear to the machine. The faster you hover backward, the more the tail wants to swap ends with the front. Of course, it's the pilot's job to keep this from happening and this requires continuous pedal corrections to keep the craft moving straight.

Actually, this is not as difficult as it sounds because you are able to see very quickly when the helicopter starts to yaw to one side or the other as you hover backward. It just takes some time to learn how to coordinate your pedal inputs to stop the helicopter from yawing so that you hover backward in a straight line.

Three things to be particularly aware of when hovering backward are speed, height above the ground, and the obstacles behind you (FIG. 6-5). If you keep your speed down to no faster than a brisk walk, you'll avoid a lot of problems, one of which is getting too low. Besides possibly exceeding the aft speed limit, a fast rearward speed requires a high nose-up attitude, which obviously puts the tail closer to the ground. You never want the tail to hit the ground and hovering backward is one of the worst times it can happen.

A hovering clearing turn should be done before hovering backward, unless you have some other way of ensuring that the area behind you is clear. (In larger helicopters, for example, you could have a crewman open the cabin door and check the area for you.) If there's room enough to do a clearing turn, there's probably room enough to hover sideways and hovering sideways is usually a safer choice than hovering backward because you can see where you're going.

United Technologies Sikorsky Aircraft

Fig. 6-5. *When hovering backward, always be aware of the obstacles behind you: Sikorsky SH-60F.*

Probably the most common reason for hovering backward is high wind. It's much easier to allow the wind to push you backward than to hover sideways with a crosswind. It might be necessary to hover backward if the wind speed is so great that turning the tail into the wind would cause you to exceed the tailwind or crosswind limitations.

Sideways hovering, like rearward hovering, requires dazzling footwork to do it correctly. The main rotor readily accepts the idea of moving to one side or the other, but the fuselage and tail rotor are reluctant to come along. It's not too hard to figure why: The fuselage and tail present a large surface area toward the relative wind. The other problem is weather vaning, again. As soon as you turn the nose of a helicopter away from the wind (and by moving sideways you are, in effect, creating a relative crosswind), the tail wants to whip around and point the nose back into the wind.

When hovering sideways, you are working against two forces: surface area resistance against the wind and weather vaning. The faster you hover sideways, the greater these forces. The wind resistance of the fuselage will cause the helicopter to roll, or lean, toward the direction of movement. Hover sideways to the left and the helicopter leans to the left. Hover to the right and it leans right. It's not uncomfortable, but it is noticeable.

Pedal action counteracts the weather vaning. Moving left, the nose wants to turn to the left; therefore, right pedal is required to keep the nose pointing straight ahead. Moving right, the nose wants to swing right, so left pedal is needed. However, because the amount of lift produced by the tail rotor is limited, it's possible to "run out of pedal" when hovering sideways (or in a crosswind, which is essentially the same thing). You'll know you've run out of pedal when you push one of the pedals all the way to the forward stop and the helicopter keeps turning the opposite direction.

Because some left pedal is already required in a no-wind hover, you'll run out of left pedal at a lower speed than you'll run out of right pedal. This means you can hover faster to the left than to the right and hold your heading in a stronger left crosswind than right crosswind. The situation is not too critical if you run out of one pedal or the other (unless you happen to be hovering in very tight quarters), because the helicopter will simply turn itself into the wind until full left or right pedal is enough to counteract the weather vaning of the tail. You might end up hovering sideways with the nose cocked to one side, but this is not dangerous if you still have control of the situation and are clear of obstacles.

IN GROUND AND OUT OF GROUND EFFECT

Ground effect is the term given to the cushion of air that builds up between the rotor system of a hovering helicopter and the ground. It occurs because the rotor blades displace air downward faster than the air can escape from beneath the helicopter (FIG. 6-6). Ground effect is one of the few free lunches in aviation because it helps support the helicopter in a hover—decreasing the power required to hover—and doesn't extract anything in return.

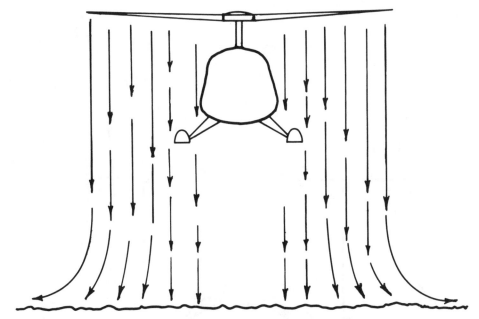

Fig. 6-6. *Air flow pattern of a helicopter hovering in ground effect.*

To hover in ground effect, the helicopter must be no higher than about one-half the diameter of the main rotor blades. The wind must also be light because any appreciable amount of wind will simply blow the descending air out from under the helicopter.

On the other hand, wind generally improves the performance of a helicopter; an increase in wind speed decreases the power required to hover. This creates an interesting hiccup on charts that prescribe power required to hover.

In a no-wind condition and at a low hover, the helicopter is in ground effect and needs a certain amount of power to hover. As the wind increases, the cushion of air moves back and away from under the helicopter, depriving it of its positive effect, but the wind is not yet strong enough to make up the difference; therefore, with about seven to 10 knots of wind, the power required to hover actually goes up. As the wind speed increases, power required goes down.

A helicopter is said to be hovering out of ground effect whenever it hovers without the benefit of the ground cushion (FIG. 6-7). Hover heights greater than one-half the rotor diameter are out of ground effect. Some surfaces, such as tall grass, steep slopes, and very rough terrain, also tend to dissipate the ground cushion and will cause the helicopter to be hovering out of ground effect at a lower height than normal (FIG. 6-8).

Ground effect cannot always be used to advantage, but when it's there it's like getting an extra bonus of power.

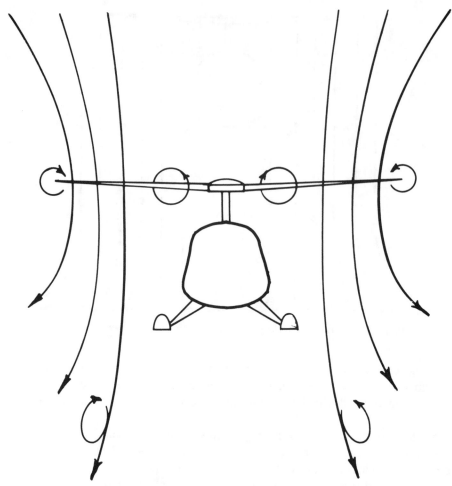

Fig. 6-7. *Air flow pattern of a helicopter hovering out of ground effect.*

Now you know everything about hovering to get started, but knowing is different from doing and no matter how much you read and study, when it gets right down to it, hovering can only be learned by doing.

Hovering is going to be tough at first. You'll become frustrated and disgusted with yourself. You'll think it's impossible and want to give up. Don't. Try, try again. It will come. With practice it will get better and eventually it will be as natural to you as riding a bicycle is now.

If it's any consolation, just remember it's something that even taxes the skill of professional pilots, particularly when checking out on a new machine. I spent over an hour during one training flight just watching and helping a Japanese Defense Force major practice takeoffs to a hover and landings from a hover in a Super Puma. He

Westland Helicopters

Fig. 6-8. *Hovering over tall grass reduces the effectiveness of ground effect: Westland Battlefield Lynx 3K.*

knew he would be flying VIPs soon and he wanted to be able to do these maneuvers perfectly. After that hour he still wasn't doing them perfectly, but they were much better than when we started. As Calvin Coolidge said, "Nothing takes the place of persistence."

Speaking of takeoffs and landings, they happen to be the subject of the next chapter.

7
More Basic Maneuvers

The trick, Fletcher, is that we are trying to overcome our limitations in
order, patiently. We don't tackle flying through rock until a little later in the
program.

Jonathan Livingston Seagull
by Richard Bach

WHAT HAVE YOU LEARNED TO DO SO FAR? HOW TO FLY STRAIGHT AND
level and how to do level turns, climbs, descents, and climbing and descending
turns. You've done all the basic hovering maneuvers. Before you do your first solo,
there are only three more things you need to learn: how to take off, land, and autorotate.

It might seem odd that learning to take off and land are the last of the basic
maneuvers you should learn, but there's a proven method of progression here. The
"building block" approach to flight instruction is the mainstay of most flight instructors worth their salt.

If your instructor follows the same order as this book, by the time you reach the
lesson on takeoffs and landings, you'll have all the prerequisites you need to do them
easily. Plus, you'll have the added bonus of having seen your instructor demonstrate
several takeoffs and landings for you. You're well prepared to do them yourself.

The only thing you haven't done is transition through translational lift airspeed.

You've flown above translational lift airspeed and you've hovered below it. There's nothing magical or difficult about flying through it; you just haven't done it yet.

TAKEOFFS

The helicopter's ability to hover creates an interesting semantic problem with the term takeoff. When does a helicopter actually take off?

Strictly and aerodynamically speaking, when you lift a helicopter into a hover, you are taking off. The FAA says you should log flight time from the instant the wheels or skids leave the ground, which is as it should be. From a maintenance point of view, everything that can move is moving, all the rotating bits and pieces are being stressed, and the clock is running on the time-change components. There should be no doubt that the helicopter has taken off.

But there is doubt.

Let's put a helicopter at an airport with a control tower (FIG. 7-1). A lot of traffic at this particular airport requires permission from ground control before engine start. You get clearance to start and you're ready to go. Because of the proximity of other aircraft, the tower doesn't want you to depart from the parking ramp, but rather from a nearby taxiway. You are cleared to taxi.

Fig. 7-1. *When does a helicopter take off? It depends: Oakland Police Department MD 500E.*

Enter ambiguity. If you are flying a helicopter with wheels, you can ground taxi just like an airplane, and the tower will expect you to do this. If, however, your helicopter has skids, you're going to have to hover taxi. Of course, it's possible to hover taxi a wheeled-helicopter, too, but this is normally only done when special conditions make it more practical or safer than ground taxiing, for example, when the ramp or taxiways are ice-covered.

You can ground taxi a helicopter with skids but it's neither good for the skids nor for the ground surface. Usually, it's better to hover taxi. At most airports, you must specifically ask the tower for permission to "hover taxi" or "air taxi," not just "taxi," because of the hazards of rotor wash on other aircraft. If you request only "to taxi," the controller might not realize you want to hover taxi and might unknowingly clear you to ground taxi too close to another aircraft.

So, with your skid-equipped helicopter, you hover taxi from the parking ramp to the taxiway. Have you taken off? The answer is yes and no.

It's "yes" for the purposes of your logbook and the maintenance records because you have taken off and can start logging flight time. It's "no" as far as the tower is concerned. All you've done is taxi to your takeoff position and, despite the fact there was air between your skids and the ground, the tower will still consider you not yet airborne. The tower wouldn't consider that you have taken off until after they have given you a takeoff clearance and you depart from your takeoff position on the taxiway.

This whole discussion might seem somewhat esoteric, but it's important to realize the distinction. It's just one of numerous things about helicopters that can be confusing. You see, by the time helicopters arrived en masse on the scene, aviation rules and regulations had already been written for airplanes. Much of what was on the books didn't apply to helicopters or meant something different. Over the years most of these ambiguities have been corrected, but many things that still apply to airplanes are confusing when applied to helicopters.

How will this affect you?

Let's say you're sitting in your helicopter, on the ground, rotors turning, at an uncontrolled airport (without an operating control tower) and your instructor says, "Okay, go ahead and take off." Does he mean take off to a hover and don't do anything but hover? Or, does he mean take off and depart the airport?

Granted, in most cases you'll probably know what your instructor wants you to do by the situation at hand, but what if you're not sure? Well, the thing to do is ask him, and don't be shy about it. Just because what your instructor says is ambiguous, doesn't mean you have to sit there silent and not know what he really wants you to do.

All right, I'm on my soapbox again. My point is that small misunderstandings and ambiguities, like the one above, often play large roles in avoidable accidents. For example, there's the true story about a twin-engine Air Force helicopter that lost one engine shortly after takeoff.

The aircraft commander shouted "takeoff power" to the copilot, meaning "move the throttle controls to the takeoff power setting" so that he would get full power on the remaining engine as he tried to gain altitude. The copilot misinterpreted the command to be, take "off" power, meaning "remove the power from both engines," which he thought could be a logical command if the aircraft commander figured they were about to crash anyway. And, of course, after the copilot shut down the remaining good engine, they did crash. A simple misunderstanding that had tragic consequences.

In any cockpit, nip misunderstandings in the bud before they blossom into full-blown accidents. Back to takeoffs . . . .

Normal Takeoff from a Hover

The most common takeoff procedure in a helicopter is from the ground to a hover and then from the hover to forward flight (FIG. 7-2). You can go directly from the ground to forward flight, without pausing in a hover, but this is usually unnecessary. Only in blowing sand, dust, snow, or other visibility-restricting matter do you really need to do it. Usually you can blow most of the junk away when you pull collective to become light on the skids (simply wait in this position until the rotor wash cleans up the area for you), so the need for a no-hover takeoff doesn't happen very often.

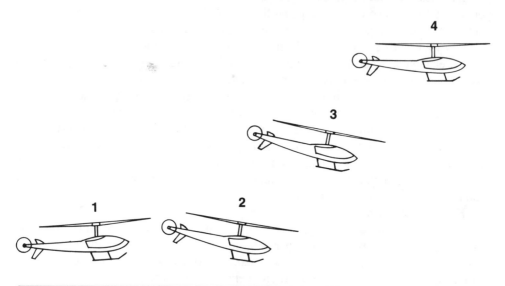

Fig. 7-2. *Normal takeoff from a hover: (1) Do hover checks at normal hovering height. (2) Ease the cyclic forward and increase the collective to prevent settling to the surface. Use the pedals to control heading. (3) Accelerate through translational lift, holding the nose-down attitude until airspeed approaches normal climb speed. (4) Raise the nose to maintain normal climb speed.*

There's a very good reason to hover for a few seconds before departing; it's one of the major advantages helicopters have over airplanes.

An airplane can't be absolutely positive about several concerns until airborne. First, takeoff is a probable time for an engine malfunction or failure, and any engine problem when you're low and slow in an airplane is a potentially dangerous situation.

Takeoff is also a critical time for the controls and control surfaces. You can and, of course, should check the controls before starting the takeoff roll, but you can never be 100 percent certain they'll work properly when the air starts flowing over the wings.

Finally, a center of gravity problem might not become evident until the airplane is off the ground and flying.

A hovering helicopter allows you to check all these things before getting too high off the ground or gaining a lot of speed. If everything works in a hover, that's a darn good indication it will all work properly in forward flight. You can check these concerns when you're still light on the skids, before you lift off the ground. If the engine seems anemic or one of the controls is binding, just lower the collective and shut the thing down.

Make it a point to hover for a moment before making a departure, whenever you can. If you don't have enough power to hover, you shouldn't take off, except under extraordinary circumstances.

The first step, then, in a normal takeoff is to lift up into a hover, which was explained in chapter 6.

Perform a *hover check* (FIG. 7-3). The hover check is your last line of defense in the never-ending battle against mechanical failures and mental lapses. One key to becoming a safer pilot is to catch lapses before they create accidents. The hover check varies from helicopter type to type, but the goal is the same. You want to make certain everything is working as it should.

Helikopter Service Aerospatiale AS 332

Fig. 7-3. *The hover check is your last chance to ensure that all the systems are operating properly.*

Check the power gauge to be sure you're using a proper amount. What's a proper amount? From experience you'll learn an amount that's normal for your type of helicopter. If the gauge tells you something else, an indication that's way too high, for example, there should be a logical explanation for it. Perhaps it's a very hot day and you're carrying a heavy load. Power required to hover would be higher than normal in such conditions. But if you can't figure out a good reason for an unusual power figure, be very suspicious. You could have a malfunctioning instrument or a problem with the engine. In both cases, have a mechanic check it out.

Check the other engine and system instruments. Usually it's sufficient to ensure that all the needles are in the green arcs: "Everything's in the green." However, like power indications, you'll soon learn what are normal temperature and pressure readings. Perhaps oil pressure is always on the low side of the green arc, but today it's nudging the top of the green. It's still in the green, but it might be (and probably is) an early indication of a problem.

Check that all the warning lights are out. You'd be surprised how easy it is to miss a warning light when you're concentrating on something else, like lifting into a stable hover. Now's a good time to check that nothing happened when you had your attention outside the cockpit.

Check the position of critical switches or handles peculiar to your type of helicopter. For example, in a wheeled helicopter, check that the parking brake is off. In helicopters with flight control systems or autopilots, check that all the switches are properly positioned.

Finally, check the cockpit controls. Most of the time, if there is a problem with the controls, you'll notice it as you lift into a hover; however, the amount of control displacement required to lift off is relatively small. Pilots have been able to take off to perfectly good, stable hovers with tail rotor or cyclic control locks installed. (It's rather difficult to not notice that the collective is locked down. You won't be able to lift it.) So double-check that the controls are working properly by displacing them slightly and noting the reaction: cyclic—left, right, forward, aft; pedals—left, right.

Your instructor will probably have a litany for the hover check. If he doesn't, devise one yourself. An easy way to remember all the items is to start at one place in the cockpit, for example the top of the control panel, and move in a line until you hit all the items you want to check.

I recall that the hover check in the Aerospatiale Super Puma was straightforward: "Power checked, no warning lights, instruments in the green, autopilot on, nose wheel locked, parking brake off, controls checked." That covered in just a few seconds a cockpit with more than 200 lights and switches.

The last item in every hover check should be to visually clear the area to your front, sides, and overhead. You never know when a person or truck or other aircraft might wander into your path and, as a helicopter pilot, you'll have to watch out for a lot more things than the average airplane pilot bound to runways.

It's time to depart.

From a stable hover, ease the cyclic forward very slightly. This will tip the rotor disc forward, cause the nose to dip slightly, and start you moving. It will also cause you to descend a bit because now you're using part of the total lift vector as forward thrust, whereas before it was only being used to counter the weight of the machine. To avoid descending, increase collective a small amount and simultaneously increase throttle to maintain rotor rpm. Of course, when you increase collective and throttle, you also increase torque, so you'll need left pedal pressure to counter the fuselage's tendency to yaw right.

You might have noticed this is the same material from the section in the previous chapter that described hovering forward, sideways, and backward. The next paragraph says:

As soon as you start to move forward, ease the cyclic back to it's stable hover position. If you don't, you'll keep accelerating and eventually pass through translational lift and enter forward flight.

Ahhhh, but now you want to enter forward flight, so don't ease back on the cyclic. Keep it pushed forward. Ease it farther and farther forward as airspeed increases and when you do pass through translational lift, you'll need a definite forward input on the cyclic to keep the nose from rising too much. Allow the helicopter to accelerate to best-climb-airspeed and then adjust the cyclic position to hold that speed (FIG. 7-4).

Fig. 7-4. *After moving through translational lift, additional forward cyclic might be needed to keep enough nose-down attitude to continue the acceleration to normal climb speed: Bell 222.*

Increased collective pitch will determine rate of climb. You can actually take off without increasing collective above that required to hover. All that is required is very gentle forward cyclic inputs to get airspeed increasing up to translational lift. Once you pass translational lift, you're home free.

The added lift means less power is required to maintain the same airspeed and if you don't reduce the collective, you'll have a reserve of power that will give you a climb. Before you get to translational lift, however, you'll be trading height for airspeed and might end up accelerating with the landing gear only inches from the ground. It's a very delicate maneuver and a good one to practice as you gain proficiency. For the time being, add some collective pitch.

How much to add will depend on conditions and the type of machine you're flying. Again, your instructor should clue you in with a rule of thumb for proper collective pitch. The more power you use, the faster you'll gain altitude.

The minimum takeoff power is hover power, the maximum amount you can use is whatever the engine can provide, subject to its limitations. It usually isn't necessary to use maximum available power for every takeoff. Unless you need all the power you can get to clear obstacles or gain altitude quickly, it's better for the engine if you don't pull max power every time. On the other hand, don't get too conservative and not use enough power to make a decent climb. Use what you need to stay safe.

The pedals provide heading control. Because you're already proficient in hovering, straight and level, and climbs and descents, holding a heading should not be a problem if you take off into the wind. Crosswind takeoffs require some explanation.

You should strive to take off into the wind, but sometimes this just isn't possible. It's permissible to take off into a crosswind with the helicopter's nose cocked to one side in a crab, but it's usually preferable to take off in a slip. If the wind is very strong, you might have no choice but to take off in a crab.

Flying in a slip, which is uncoordinated flight, requires that the cyclic be held into the wind and the pedals used to keep the heading straight along the takeoff path. The goal is to have the fuselage and the ground track aligned so that in the event an immediate landing is necessary the chance of touching down skewed to one side is minimized.

Once you've obtained translational lift and a safe altitude, allow the helicopter to enter a crab into the wind while maintaining the same ground track. The helicopter will be in coordinated flight again.

Continue the climb until you reach the desired cruising altitude and level off as described in chapter 5.

Takeoff from the Surface

The main reason for taking off from the surface without hovering is because of the possibility of blowing snow, dust, or other matter reducing visibility to zero while in a hover. In most cases, however, you should be able to blow away most of the obscuring material by becoming light on the skids and holding the helicopter there for a bit longer than normal. If you can hover, and usually you can, you should hover to check aircraft systems.

If you must take off from the surface, do a modified hover check while the helicopter is light on the skids. Expect to create a cloud of dust or snow as you do this, so find several hover references before you start. Then lower the collective again, being careful to keep the cyclic and pedal in the same positions. You want to jump off the ground quickly to avoid the cloud of material, so it doesn't make sense to start off with the cloud already there.

You are now set to make a fast departure. Pull the collective up smoothly to maximum power while maintaining rotor rpm with the throttle. Only minor corrections should be needed with cyclic to keep the helicopter in a level attitude. Left pedal is needed to counteract torque and hold the heading. The liftoff should be almost vertical, but not quite.

You want to have some forward motion so that airspeed will increase. The helicopter will create a cloud with the rotorwash and you'll very quickly pop out in front of and above it. As soon as you do, lower the nose to attain normal climb speed. From then on the maneuver is just like any other takeoff.

Running Takeoff

Running takeoffs and landings are more appropriate to wheeled helicopters than those with skid-type landing gear. Skids wear out fast if you continually slide them over hard surfaces and they tear up soft surfaces. The main reason for making a running takeoff is to avoid having to use the power required to hover.

Every helicopter, whether wheeled or skidded, will have a running takeoff and landing groundspeed limitation. Commonly it's close to the translational lift airspeed, or slightly above; thus, in a no-wind situation, you can make a running landing while still flying at translational lift airspeed. Conversely, you could make a running takeoff and attain translational lift speed while still on the ground.

Running takeoffs increase the operational envelope of the helicopter because they allow you to take off or land when you don't have enough power to hover. You shouldn't make a habit of this, for reasons I explained before, but it does give you the option.

In a wheeled helicopter, a running takeoff is similar to a takeoff in an airplane. After doing as much of the hover check as you can, you increase collective to a power equal to at least one-half the power required to hover and apply forward cyclic. This tilts the rotor disc forward and starts the helicopter rolling. Pedals are used to maintain heading as airspeed increases.

After passing translational lift airspeed there really is nothing more to be gained by staying on the ground any longer. Simultaneously, increase collective pitch to lift the helicopter off the runway and apply cyclic as needed (fore or aft) to attain a climb attitude. Certain helicopters tend to tuck their nose as they lift off from a running takeoff; others rise level off the ground. From then on, it's a normal climb.

Running takeoffs in skidded helicopters are done the same way, although you definitely don't want to dally on the ground too long (FIG. 7-5). To repeat, a running takeoff in a helicopter with skids is not a normal maneuver. The only reason to do it is if you don't have enough power to hover. If you don't have enough power to hover it's because the helicopter is too heavy for the atmospheric conditions.

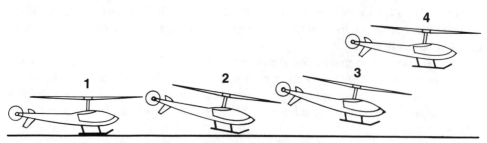

Fig. 7-5. *Running or high-altitude takeoff: (1) Set power to just below hover power and ease cyclic forward to start slow acceleration. Maintain heading with pedals. (2) After passing translational lift airspeed and before reaching the landing gear groundspeed limitation, ease back on the cyclic to become airborne. (3) Maintain a level attitude over the surface to accelerate to normal climb speed. (4) Raise the nose to maintain normal climb speed.*

The Bell Helicopter Textron training guide for the Bell 206, *Flying Your Bell Model 206 JetRanger,* comments about running takeoffs. Consider it applicable to all helicopters:

> No flight should be attempted without the capability of at least being able to hover momentarily. If the helicopter is loaded to gross, some possibility exists that it has been loaded in a manner to exceed the center of gravity limitations, and insufficient cyclic control to allow safe flight might be present. If you are unable to get clear of the ground enough to check your hover cyclic position before exceeding maximum power limits, you'd be wise to off-load part of what you intend to carry and make two trips. No one other than the pilot should make this decision. It might not be your helicopter, but it's your life and professional standing as a pilot that is at stake.

If you have enough power to hover, make a takeoff from a hover. If you don't have enough power to hover, reduce the gross weight so you get enough power. If you can't hover and you can't reduce the weight, don't make a running takeoff. The performance of the helicopter is unpredictable and there is no way of knowing if you have enough power available to make a safe running takeoff.

APPROACH AND LANDING

There are three different landing approaches: to a hover, to a running landing, and to the ground. All three begin the same way, with a normal descent you learned in chapter 5. It's the way they end up that makes them different.

Normal Approach to a Hover

An approach to a hover is the one you'll make the most often in a helicopter (FIG. 7-6). The difficulty of the maneuver derives from the fact that you must lose altitude and airspeed to transition from forward flight hundreds of feet above the ground to a hover only a few feet over one particular spot.

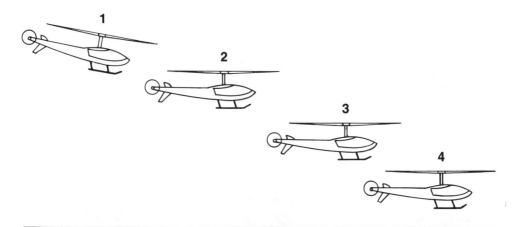

Fig. 7-6. *Normal approach to a hover: (1) Line up on final with normal approach airspeed, power, and rpm. (2) Lower the collective to start the descent. Adjust the cyclic to maintain airspeed. Use the collective to maintain a constant sight picture (angle of descent). Pedals control the heading. (3) At the manufacturer's recommended altitude, apply aft cyclic to decrease groundspeed and start the landing flare. Increase the collective as the airspeed drops below translational lift so that the rate of descent does not increase. (4) Use forward cyclic and collective as needed to stop in a level hover over the landing spot.*

Don't kid yourself into thinking it's easy. It's the most difficult of the basic maneuvers. It amounts to three problems that require simultaneous control: rate of closure, rate of descent, and heading control. But as I mentioned in the beginning of the chapter, you have all the skills you need to do it.

Start the approach by lowering the collective pitch the amount required to descend at an angle of approximately 15 degrees on final approach. As collective pitch is lowered, increase right pedal pressure to compensate for the change in torque and maintain heading. Adjust the throttle to maintain rotor rpm. Decelerate to landing approach speed and adjust the cyclic to maintain this speed.

The angle of descent is controlled by collective pitch; airspeed is controlled by cyclic; coordination of all controls is required continuously. Maintain a constant sight picture of the landing area to keep a constant angle of descent (FIG. 7-7).

Because you want to end up with zero groundspeed, at some point on the approach you will have to begin to reduce airspeed. Deciding when to do this is probably the most difficult part of the approach because the point will vary with the wind. The goal should be to maintain the approach angle without shallowing out or getting too steep, and to have a steady rate of deceleration and descent. Even experienced pilots don't get it right every time. Fortunately, if you don't start to decelerate at precisely the right point you can still make an acceptable approach by varying your rate of descent and rate of deceleration.

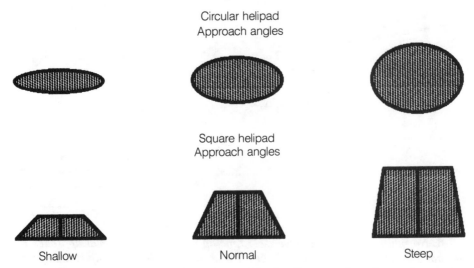

Circular helipad
Approach angles

Square helipad
Approach angles

Shallow   Normal   Steep

Fig. 7-7. *Maintain a constant sight picture of the landing area in order to fly a constant descent angle.*

As you descend closer to the ground, apparent groundspeed will appear to increase. Resist the temptation to slow down too soon. One position you want to avoid is being too high and too slow because that might make it difficult to make a good autorotation if the engine fails. The other position to avoid is too fast and too low because you might not have enough power to stop the descent or might flare excessively and hit the tail.

Generally speaking, you should decelerate through translational lift close enough to your intended landing point so that you can make the spot if the engine fails.

Remember, loss of translational lift means an increase in power required. Once you lose it, you enter the realm of hovering flight, which, as you know, requires more power than forward flight. So, as you come through translational lift, you'll have to be increasing the collective to compensate for the loss of lift (FIG. 7-8). If you fly the approach properly, you shouldn't need to pull any more power than that required to hover. If you find yourself sinking toward the ground, though, don't be afraid to pull whatever power you need to keep from touching down.

Now that you're hovering, you'll feel like you're back in familiar territory. Coordinate the controls as you did before. If you are hovering over your intended landing spot, simply land as you would from a normal hover.

It is very common to undershoot or overshoot your landing spot on your first attempts at landing approaches. Don't worry, it will come with time. Undershooting usually isn't too bad because you can simply hover forward in most cases. Overshooting can be more of a problem, particularly if the landing area is small or if you try to salvage the approach by making a big flare to lose airspeed. There is no shame in making a go-around if you overshoot your approach. In such cases, discretion is the better part of valor.

Fig. 7-8. *As a helicopter decelerates below translational lift, additional power is required to prevent settling to the surface before reaching the landing spot: Boeing 234, North Sea.*

Normal Approach to the Surface

If you were to rate landings by difficulty, this type would come out on top. When done properly, however, it's quite a handy maneuver to have in situations where you don't want to, or can't hover, and the surface is too rough for a running landing.

The main reason you would want to make an approach directly to the ground without stopping in a hover is the same as for wanting to take off from the surface without hovering: possibility of snow, dust, or other materials restricting visibility in a hover. Another reason is that you simply do not have enough power to hover. A partial power failure in a single-engine helicopter or the failure of one engine in a twin-engine helicopter are good examples. Most twin-engine helicopters do not have single-engine hover capability except under very favorable conditions (low gross weight, low temperature, strong headwind).

A normal approach to the surface is accomplished just like a normal approach to hover, except that instead of stopping in a hover, you take the machine all the way to the ground in one fluid motion. The tricky parts are making the motion fluid, not landing too hard, and avoiding stopping in a hover, if you do have enough power to hover, because you were too worried about landing hard.

Ideally the rotor wash is going to blow away all the snow or dust when closer to the ground. Realistically the depth and density of the particles plays a part in the scenario.

In Alaska, we found there were two ways to land on snow-covered terrain. The first was to stop in a high hover, 50 to 100 feet above the ground, and slowly descend while blowing the snow away. This worked fine if there wasn't a lot of snow or if there was a thin layer of light snow over a crust. If there was too much snow or we didn't have the time or power available to blow the snow away, we made an approach to the surface.

As you approach the spot in such conditions, the helicopter will create a rolling cloud of dust or snow behind it. The nearer you get to the surface and the slower you go, the closer this cloud comes. If you look behind you, you can see it approaching the aircraft. Just before you come to a stop, it closes in from both the sides, then surrounds you completely.

Visibility will only be a few feet; therefore, it's very important to pick out a landing reference before the dust or snow cloud envelopes the entire machine. Find something that will stand out, such as a black stump or rock in a snow-covered area. The more references you can find, the better, but if you only have one, try to land so it's located at about a 45-degree angle from the nose of the aircraft and no more than a few feet away.

The key to a successful approach to the surface is to know how much power to pull. You should have in your mind a power setting that will give you hover power. If you haven't burned off a lot of fuel and your landing site is about the same altitude as your takeoff site, hover power will be about the same. As you approach the spot, keep in mind that this power setting will cause the helicopter to stop in a hover. You can pull it to slow down the descent, but don't hold it too long. You need a slightly lower power setting to allow the helicopter to settle to the ground.

Don't forget about ground effect. As you descend from a normal hover height closer and closer to the surface, ground effect increases; therefore, you have to progressively reduce collective pitch. Don't slam it down, but don't be too timid about it either.

When the obscuring cloud engulfs you, you should be on the ground or just about to touch down with the hover reference in sight.

The application of collective pitch could be summed up as follows. Start the approach with a decrease in collective pitch. As the helicopter slows below translational lift airspeed, gradually increase collective pitch. Maximum collective pitch should be reached just before the helicopter passes through normal hover height; however, the maximum amount of pitch used should not exceed that required to hover. Hold maximum collective pitch for a second as the machine settles to a level attitude, then lower it carefully to allow the helicopter to continue down to the ground. Once you are safely on the surface, lower the collective all the way down.

Running Landing

Running landings are the easiest landings to make, particularly in a wheeled heli-copter; such landings are easier to make in a helicopter than they are in an airplane. Some helicopter operators instruct their pilots to make running landings whenever they can as a matter of routine in order to avoid the high power settings required by hovering. In twin-engine helicopters, running landings might be necessary in the event of a single-engine failure, so the maneuver is practiced often in training.

An approach to a running landing is normally shallower than a normal approach to a hover (FIG. 7-9). Start with the same approach speed you use with a normal approach. The main difference is that you do not have to slow your speed as much as you do with an approach to a hover; therefore, you avoid the most difficult part of an approach to a hover, namely the transition from forward flight to hovering flight. You also use less power and fewer control inputs.

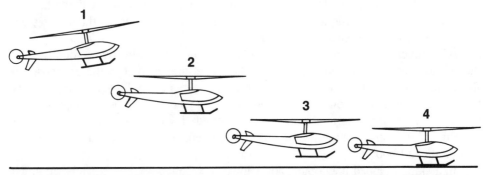

Fig. 7-9. *Running landing: (1) Line up on final with normal approach airspeed, power, and rpm. (2) At a point slightly earlier on final than with a normal landing, lower the collective to start a shallower-than-normal descent. Adjust the cyclic to maintain airspeed. Use the collective to maintain a constant sight picture (angle of descent). Pedals control the heading. (3) Maintain normal descent airspeed until just short of the end of the land-ing area, then gradually reduce airspeed to translational lift airspeed. (4) Allow the heli-copter to settle to the landing area in a level or very slightly nose-high attitude. Use a slight increase in collective to cushion the landing. Center the cyclic and carefully lower the collective. Apply brakes, if wheeled landing gear.*

Basically, you fly straight toward your touchdown spot with a constant airspeed. At about the same height (50 to 100 feet, depending on type) you start the deceleration on a normal approach, ease back on the cyclic to raise the nose. This will slow air-speed and initially reduce the rate of descent. Because you're on the back-side of the power curve, a slower airspeed at the same power setting will result in a higher rate of descent, but this doesn't take effect immediately. As you descend closer to the ground, level the aircraft and add collective to maintain the reduced rate of descent.

The airspeed should be just above translational lift speed as the helicopter descends into ground effect. As translational lift dissipates, ground effect takes hold and a very slight increase in collective, if any at all, will be needed to cushion the landing.

The surface of the landing area and the flight manual limitation on the landing gear will dictate the touchdown speed. Generally speaking, you can safely land a wheeled helicopter at a higher groundspeed than you can land a skid-equipped helicopter. To obtain a lower speed, hold the helicopter off the ground by progressively adding more power with the collective and easing the cyclic aft. Don't do this too low to the ground or you might ding the tail. If you know you have to land with a low touchdown speed, it's better to get rid of the excess airspeed before you get into the ground cushion.

Your goal should be to land as level and as straight as possible. If there is a crosswind, you might crab as long as you want on the approach, or you can slip to keep the nose aligned with the ground track, but you must touchdown in a slip with the fuselage straight with the line of flight. Touching down skewed could damage the landing gear or even cause the helicopter to roll over.

After touching down in a wheeled helicopter, lower the collective smoothly to put the aircraft firmly on the ground and use the wheel brakes to slow the groundspeed. Do not use aft cyclic to try to slow your speed. Believe me, you'll want to try, because that's the way you slow down in hovering flight. As you descend on the approach, tell yourself, "I will not use aft cyclic on the ground. I will not use aft cyclic on the ground." If anything, use a little forward cyclic to ensure that the main rotor does not contact the tail boom.

In a skid-equipped helicopter, don't lower the collective immediately on touchdown. Skids have a much higher coefficient of friction with the surface than wheels and you risk a sudden stop if you make the aircraft "heavy on the skids" when it still has too much forward speed. Stop too quickly and the helicopter might nose over. If you feel you must obtain more braking action, lower the collective very cautiously.

WORDS ABOUT WIND

Helicopters are versatile vehicles that can do many things, but they operate better under some conditions than others. With respect to wind, you should always strive to hover, take off, and land into the wind whenever possible. Taking off and landing downwind—with your tail into the wind—is asking for trouble.

With the wind pushing from behind, it takes more power to hover than it does in a no-wind condition. It's also harder to hold a steady hover because the front and back want to swap positions. A helicopter is just like a big weather vane in this respect. Compare the distance between the nose and the rotor mast and the rotor mast and the tail and you'll see why this is so.

Another concern with the tail into the wind is the air flow coming off the tail rotor and meeting the main rotor. Helicopters are designed so that both rotors operate in clean air most of the time because rotor effectiveness is lost if the air is turbulent. With the nose into the wind, the tail rotor is above the main rotor wash and, of course, the tail rotorwash moves to the side and rear. Put the tail into the wind and the main rotor will be running into some of the tail rotorwash at some point. It's usually not a big problem, but it does mean you have to work harder to keep the hover stable.

The transition from a hover through translational lift takes longer if you start with a tailwind. Remember, a helicopter goes through translational lift at about 15 to 25 knots of forward airspeed, not groundspeed. With a downwind hover of, say, 15 knots, you have minus 15 knots of forward airspeed. To get to translational lift airspeed, you must first restore those 15 knots and then accelerate another 15 or 25 knots, for a total airspeed increase of 30 to 40 knots.

This takes time and eats up a lot of ground. It's not a big problem if you're taking off from a long runway or open area, but if the area is confined you might have problems. What is more dangerous is that you will be moving over the ground much faster than you would during a no-wind takeoff. Compare your groundspeed if you take off into that 15-knot wind.

Now, in a hover, you have a positive airspeed of 15 knots and you use less power than in a no-wind hover. With the nose into the wind, the streamline design of the helicopter makes yaw control easier because the machine wants to point this way. A small nudge forward on the cyclic tilts the rotor disc forward and starts the machine moving slowly over the ground. Airspeed increases quickly and, before you've moved very far, you've reached translational lift. The airspeed indicator shows 25 knots, but the groundspeed is only 10 knots.

An upwind takeoff is much safer if you have a mechanical problem and have to put the machine on the ground. To make a zero-groundspeed touchdown, which obviously has the lowest potential for injury or damage, you wouldn't have to decelerate the machine nearly as much as you would if you take off downwind.

A comparison between landing upwind or downwind is similar. During a normal no-wind landing, the task is to descend from final approach altitude while decelerating from approach speed to zero groundspeed (a hover). Total change in airspeed is about 50 to 70 knots, depending on aircraft type.

With a headwind of 15 knots and a normal approach speed of 60 knots, the task is somewhat easier because now you only have to decelerate 45 knots in order to stop in a hover. Fewer control inputs are required all around. With 15 knots on the nose in a hover, you need less power: not as much collective and throttle application. Because the wind helps slow the machine down, you don't need as much aft cyclic. With fewer collective and throttle movements and the tendency for the helicopter to weather vane into the wind, you need fewer tail rotor pedal inputs. Finally, you don't need as much real estate to get the machine slowed down.

Landing downwind increases your problems. Now you need to decelerate from 60 knots to minus 15 knots, a total of 75 knots, and then stop in a downwind hover with all its attendant problems. The number of control changes increases dramatically. You need to reduce collective and throttle more and apply a hefty aft cyclic movement.

Yaw control becomes more and more critical as you slow down. As you get closer to a landing spot, you need a greater nose-up attitude to stop forward movement over the ground. You need a large collective application to stop the descent. The danger is a too-low height and too much nose-up and a consequent tail strike on the ground. Miscalculation of the aircraft's position and height and misjudgment of the control inputs required are common problems during downwind landings.

The problems of crosswind takeoffs and landings are like those of downwind takeoffs and landings to a lesser degree, with one exception. A strong crosswind from the right—in American-made helicopters—could cause you to run out of tail rotor control as the airspeed decreases. The reason is that the tail rotor can only produce a limited amount of lift and might not be able to produce enough lift to counteract torque effect and a right crosswind.

In a normal hover, you need some left pedal to counteract the torque of the main rotor. With a crosswind from the right, the helicopter wants to yaw to right and additional left pedal is required to hold the nose straight. With a strong wind from the right, you can push the left pedal all the way to the stop and not have enough tail rotor force to stop the machine from yawing to the right. The right yaw will continue until the helicopter aligns itself into the wind sufficiently for the full left pedal to counteract the torque of the main rotor and the remaining crosswind component.

This is not to say that downwind and crosswind landings and takeoffs are impossible and you'll never have to do them. Obstacles will sometimes require a takeoff or landing in a direction other than straight into the wind. Be aware of the hazards and be prepared to meet them.

TRAFFIC PATTERNS

The normal way to practice takeoffs and landings is in a traffic pattern. Many of the maneuvers covered are complicated; therefore, the following excerpt from the Bell Helicopter Textron training guide, *Flying Your Bell Model 206 JetRanger*, makes a good summary. Please be aware that the altitudes, airspeeds, and power settings will vary for different helicopter types.

The helicopter's special flight abilities do not require standard landing patterns. Because helicopters do not need runways, the tower may clear you to land or take off crosswind or parallel to their active runway, or from a taxiway or parking ramp, particularly at airports where airplanes are operating.

For training purposes, the pattern presented here is chosen for safety, its conformity to FAA and ATC procedures, and its training value. A plan view of a normal left traffic rectangular pattern is shown in FIG. 7-10. For right traffic, use a mirror image of the left traffic pattern. Any part of the pattern can be varied to suit terrain or other conditions, but it's good to first learn to fly the standard pattern with precision before making variations.

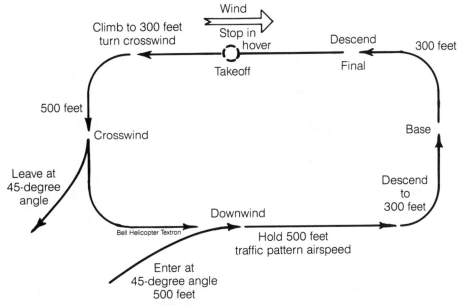

Fig. 7-10. *Normal left-hand traffic pattern as taught for the Bell 206. A right-hand traffic pattern is a mirror image.*

Normal Takeoff or Departure from a Hover

If the terrain permits, the takeoff should be made over a smooth, unobstructed area. A look at the height/velocity diagram (*see* page 165) in your helicopter flight manual will readily explain the reason for this. Stated simply, a safe landing in case of engine failure is doubtful from altitudes between 50 feet and 200 feet if your airspeed is less than 40 knots; therefore, plan your takeoff to avoid this altitude/airspeed area of doubtful safety.

Hover into the takeoff position and line up on your takeoff leg. Look ahead and familiarize yourself with any obstructions or favorable areas that will influence your proposed ground track. When you are satisfied that your takeoff leg is safe, gently lower the nose and begin hovering forward along your ground track.

A slight settling of the helicopter will occur as you begin to move forward because a portion of the power that has been producing lift is now being used for thrust. To compensate, a small up-pressure on the collective will hold your altitude.

Continue hovering forward and accelerating smoothly, and at about 15 to 20 knots you will feel a considerable increase in lift and the helicopter will tend to yaw left as the main and tail rotors move into translational lift. The nose will also try to pitch up slightly. Just before reaching this condition, you will be alerted to it by a moderate vibration or shudder throughout the helicopter.

Coordinate the cyclic and pedals to maintain a slightly nose-low attitude and heading parallel to your movement over the ground. The helicopter will climb in this attitude and continue to accelerate. Gently ease the nose forward and allow the helicopter to accelerate to the best rate of climb speed (V_y). At this speed ease the nose up to a slightly nose-low attitude (two or three degrees below the horizon) and the helicopter will climb out smoothly.

During this maneuver divide your attention between the power gauge (manifold pressure for reciprocating engines or torque for turbine engines) and the engine temperature gauge to adjust climb power; also, the airspeed indicator and your attitude with relation to the horizon to adjust your climb airspeed.

Directional correction for any crosswind should consist of a slip until well clear of the ground (50 feet or above) so that your landing gear is paralleling your ground track. After reaching this altitude, enter a crab into the wind to climb out coordinated and streamlined.

Crosswind

It is best to climb to at least 300 feet before turning crosswind. Because airspeed and power setting are stabilized by this time, a simple cyclic turn puts you on the crosswind leg. Remember, some helicopters require just a "feel" of pedal in a turn.

To make the turn smooth and at a constant climb airspeed, it is best to clear your turn first, then look ahead and control your attitude by reference to the horizon. It is very easy to let your attitude wander and lose your best climb airspeed. Plan your turn to roll out level with enough crab to correct for wind drift and fly a 90-degree ground track.

Downwind

Five hundred feet is a safe altitude to turn downwind, giving you plenty of time to make a smooth turn into the wind in case you must make an autorotation because of an engine failure. As you approach 500 feet, build your speed by easing the nose down a few degrees leaving climb power in until you've reached power to cruise, about 70 to 80 percent power being enough to give you 90 to 115 knots.

This sounds deceptively simple. It will take a little practice to smooth up and coordinate this power change in a turn while holding a constant speed. If done properly, you will roll out on downwind smoothly at cruise power and airspeed with very little or no correction.

Base

For training purposes, set your base leg about a quarter mile downwind from your intended approach point so that you won't be rushed and will have an adequate amount of straight and level flight on final to plan a smooth approach.

Start a cyclic turn, then reduce power to descend to 300 feet on base. This is good practice for descending turns. Roll out with enough crab to keep the pattern square and reduce speed as you continue letting down so that you arrive at the turn to final at V_y and 300 feet altitude. Plan your turn to final so that you will be lined up with your approach point.

Final Leg and Normal Approach

To aid in your practice for a smooth, well-planned approach, you should roll out at 300 feet and V_y with about one-eighth of a mile of final to fly straight and level. This gives you time to get trimmed and look over your landing area before entering the approach. Give yourself enough room so that you won't be rushed. It is practically impossible to fly smoothly while you are hurrying.

Now remember, you're not on final for a landing to the surface. You're only going to fly down from 300 feet at V_y while decelerating and stop in a hover at three feet with zero groundspeed. Don't get all tensed up and ready for a big bunch of problems. You'll still be flying when you're through with the approach. After that, you'll descend straight down from a hover either at this point or after hovering to some other spot on the airfield.

The safest and generally most satisfactory approach is down a 10- to 20-degree approach angle. This angle gives clearance over obstacles into confined areas, yet allows you to keep your intended landing area in sight all the way down. As you arrive on final to a position where your spot is about 10 degrees below the horizon and any approach obstacles have dropped below your approach path, enter the approach by lowering the collective.

You'll have to coordinate pedal to maintain heading. Attitude should be held constant during the first half of the approach. Do not let the attitude go too far astray. Control your angle of descent with the collective. That is, if your approach path looks short, add up-pressure; if long, apply down-pressure.

At about 100 feet, begin dissipating your forward speed with the cyclic, slowly adjusting your rate of closure so that you arrive at a full stop over your approach spot. As you decelerate, you will be making a transition from translational lift through transverse flow effect to a hover. Additional power will be required with slower airspeed as the rotor begins operating below translational lift. The power difference is usually about 10 or more percent (FIG. 7-11).

The addition of power should be smooth and continuous, but only at a rate you need in order to maintain your selected line of approach. As power increases, you will need to begin applying left pedal against the increased torque. Just at termination, the nose of some helicopters has a slight tendency to pitch up, so be prepared with a little forward cyclic pressure which will allow the helicopter to coast to a three-foot hover.

This maneuver takes practice and requires "feel." It is not like landing an airplane, in the sense that you try to touch down just above a stall. You simply bring the helicopter to a hover, so relax. You'll make a few mistakes, but there's nothing critical about them if not carried to extremes. Relax and stay loose. Keep your eyes moving.

Fig. 7-11. *Bell 206L LongRanger about to enter a hover during a normal approach.*

QUICK STOPS

A *rapid deceleration* (quick stop) is primarily a coordination exercise, although you might have occasion to use it outside training, too. It's purpose is to quickly bring the helicopter to a stationary hover from forward flight.

Quick stops are like accels/decels only quicker. They're also done closer to the ground. Because of these two factors, they require a greater degree of skill, coordination, and vigilance.

Begin the exercise from a normal hover with the nose pointed directly into the wind and with a thousand feet or so of clear, flat area, such as a runway, in front of you. Start as if you were making a normal takeoff from a hover. Accelerate through translational lift and allow the helicopter to climb to about 25 to 30 feet. This height should be high enough to avoid danger to the tail rotor during the flare, but low enough to stay out of the shaded area of height-velocity chart during the entire maneuver. Level off at this height and continue to accelerate to no more than 40 knots.

Begin the quick stop by smoothly lowering the collective, simultaneously applying right pedal to maintain heading, decreasing the throttle to keep rpm within limits, and easing back on the cyclic to decelerate the helicopter. The timing of the controls' inputs must be precise and you'll understand the value of doing accels/decels before attempting quick stops. If you use too much aft cyclic or apply it too quickly, the helicopter will climb. If you use too little cyclic or apply it too slowly, the helicopter will descend.

Hold aft cyclic until the helicopter has decelerated to the speed of a brisk walk. Lower the nose slightly and, as the helicopter continues to descend, start increasing the collective to stop the helicopter at a normal hover height and zero groundspeed.

The throttle will have to be increased to maintain rpm and left pedal applied to hold the heading.

Be very careful to avoid an excessive tail-low attitude too close to the ground. Give yourself plenty of room too slow down so you won't be tempted to pull back too much on the cyclic in order to stop in a small space.

Quick stops aren't something you'll do every day in the real world, but they might come in handy. At busy airports, the tower controllers often clear helicopters to depart across an active runway. If a controller misjudges incoming or departing fixed-wing traffic, you might find it necessary to abort your takeoff by doing a quick stop in order to avoid coming too close to another aircraft.

Now that you have all the basics, it's time to do your first emergency procedure: autorotations.

8
Autorotation

On June 21, 1972, in an Aerospatiale SA-315B Lama that was lightened as much as possible, I reached an altitude of 40,814 feet, establishing a helicopter altitude record which remains today. That flight also wound up being the longest autorotation in history because the turbine died as soon as I reduced power. With a −63 °C temperature that day, the engine flamed out and could not be restarted.

Jean Boulet,
former Aerospatiale chief test pilot
Rotor & Wing International Magazine, 1991

RECALL THE FIRST MYTH IN CHAPTER 1: IF THE ENGINE QUITS, YOU'RE a goner. Also recall the first fact: You have a better chance of survival after a complete power failure in a helicopter than you do after a complete power failure in an airplane. This is true, you recall, because of the helicopter's autorotative capability. Now that you know how to do the basic flight maneuvers, it's time to talk about autorotations.

Autorotating is an emergency maneuver and not something you'll do every day. On the other hand, it's an extremely important emergency maneuver that is necessary to learn early in your training—before you do your first solo—and to stay proficient in throughout your career as a helicopter pilot.

To put your mind at ease, autorotations are not that difficult. They are easier than some of the advanced maneuvers. By the time you start to practice autorotations, you'll probably find them easier than your first attempt at hovering. They can be frightening at first because the rate of descent is greater than a normal rate of descent; you might even feel as if you're falling.

Fear of falling, the psychologists tell us, is one of two basic fears we all are born with. (Fear of loud noises is the other.) Try not to be intimidated by the high sink rate. You won't really be falling, you'll actually be gliding steeply. It's only natural for the descent rate to be high and for it to be a little scary. If you don't mind roller coasters, you won't mind autorotations at all. To be honest, I'd choose autorotating in a helicopter over a roller coaster ride any day. Autorotations are much more tame.

Engine failures in modern aircraft, fixed-wing and rotary-wing, are rare. In some 9,000 hours of flight time, I've only experienced one actual engine failure, and that was in a twin-engine helicopter. I've had to shut down an engine a few times, mainly because of a low oil pressure indication, but, again, that was in twin-engine helicopters. Had I been in a single-engine helicopter, it would have been necessary to autorotate after shutting down the engine, but at least the engine shutdown would not have come as a surprise.

I do know other helicopter pilots who have had to autorotate three or four times during their careers, so maybe I've just been lucky. Because you'll be flying single-engine helicopters during your training and perhaps later, too, you should be prepared to autorotate at any time.

An engine failure in a single-engine airplane can be quite critical, especially if it occurs close to the ground and at a low airspeed. Immediate action is required. On the other hand, if the failure occurs at a high altitude and at normal cruising speed, the situation is serious, but not as critical. The pilot could conceivably do nothing and the airplane, although descending in a glide, will continue to fly. Things don't get really critical until the airplane is closer to the ground.

An engine failure in a single-engine helicopter always requires immediate action from the pilot regardless of altitude—always (FIG. 8-1). That action, lowering the collective to decrease the pitch angle of the main rotor blades, is absolutely necessary in order to keep the rotor rpm within limits and maintain the flying capability of the helicopter. With excessive pitch, there is simply too much drag on the rotor blades and the rotor rpm will decay (decrease). If rotor rpm is allowed to decay too much, control of the helicopter will be lost and it will be impossible to get it back. The helicopter will then fall like the proverbial brick.

To repeat, the first action whenever the engine fails in a single-engine helicopter is to lower the collective, except in a low hover, which is explained in this chapter.

FOUR-STEP AIRCRAFT EMERGENCY PROCEDURE

Because we are talking about an emergency maneuver, this is a good time to introduce the "Basic Four-step Aircraft Emergency Procedure." Years ago, I ran across

McDonnell Douglas Helicopter Company

Fig. 8-1. *An engine failure in a single-engine helicopter always requires an immediate response from the pilot: McDonnell Douglas MD 500E (for illustration purposes only).*

this in an article in an aviation magazine. The author attributed the procedure to his old flight instructor. I don't know who thought of it first, but it's certainly worth repeating:

Step 1. Fly the aircraft.
Step 2. Remember step 1.
Step 3. Immediate action items.
Step 4. Checklist.

I thought this sounded so good, that when I was involved in revising the S-61 and AS332 emergency checklists for Helikopter Service of Norway, I convinced the chief pilot to print this basic checklist on the front cover, slightly revising step 1 to better fit helicopters:

Step 1. Fly the aircraft and maintain rotor rpm.
Step 2. Remember step 1.
Step 3. Immediate action items.
Step 4. Checklist.

With any emergency or unusual occurrence in any type of aircraft, your main concern is to keep the machine flying safely in the air until you can put it on the ground. Pay attention to heading, altitude, and airspeed—that's what fly the aircraft means. In a helicopter you have one more thing to remember so you can take care of the above three parameters: maintain rotor rpm.

Why is proper rotor rpm so important? It has to do with lift and airspeed and angle-of-attack. If the main rotor isn't spinning fast enough, it just can't make any lift. Neither can the tail rotor. This is even worse because you can lose yaw control while the main rotor is still producing some lift. Because the tail rotor is spinning about five times faster than the main rotor and lift varies as the square of the velocity, you'll lose yaw control rather quickly as rotor rpm decreases. That's all there is to it. Period.

PRACTICING AUTOROTATIONS

In normal flight, the engine or engines power the main transmission and the main transmission transmits the power to the main and tail rotors. If an engine quits, the freewheeling mechanism automatically uncouples the engine input to the transmission so that the transmission and the rotor blades can rotate unencumbered by the engine. This "automatic rotation" of the rotor blades after uncoupling a failed engine is the mechanical essence of autorotation.

To practice autorotations, the engine or engines could be shut down completely, but you would be burning bridges behind you. You wouldn't have an out if you made any mistake during the autorotation procedure.

Fortunately, it's possible to do autorotations with the engine or engines running and this is the normal way to practice them. A partial reduction in engine power is all that's required to slow down the engine-to-main-gear-box-input shaft enough to cause the freewheeling unit to uncouple the transmission. In the cockpit, you look for the engine rpm (N2 or Nf) needle to decrease below the rotor rpm (Nr) needle. Once the needles are split, the main rotor is rotating freely on its own and is independent of the engines.

(From here on, I'll refer only to single-engine helicopters so I don't have to keep writing "or engines." Just remember that both engines failing in a twin-engine helicopter is the same as one engine failing in a single engine helicopter.)

Your first practice autorotations should be started from a comfortable altitude, say 3,000 feet, and end with a power recovery well above the ground. The point of the exercise is to learn how to enter an autorotation, to get a feel for the helicopter in an autorotative descent, and to experience a flare and power recovery.

Just in case the engine really does quit while you're practicing autorotations, you should do them over an area that would be suitable for landing. Start in straight and level flight, into the wind, at an airspeed equal to or only a few knots above best autorotation speed.

Every helicopter has a "best glide speed" or "best autorotation speed," the airspeed that gives the slowest rate of descent in autorotation, usually between 50 and 70 knots. The glide distance can be improved by descending at a higher airspeed, up to a point, but if you use a lower speed you'll go down at a faster rate and you won't glide as far.

Start the first couple of autorotations close to the best autorotation speed so you don't have to work too hard to maintain that airspeed. You'll also need time to concentrate on everything else that's happening. As your proficiency in autorotations improves, work up to starting your practice autorotations at high cruise speeds because these will be the airspeeds you'll be flying more often.

Lower the collective all the way down, then rotate the throttle off ("power off") until engine rpm splits off from the rotor rpm. When the engine stops delivering power to the transmission, it's the air flow that's turning the rotor (FIG. 8-2). The helicopter has become a horizontal windmill and you're autorotating.

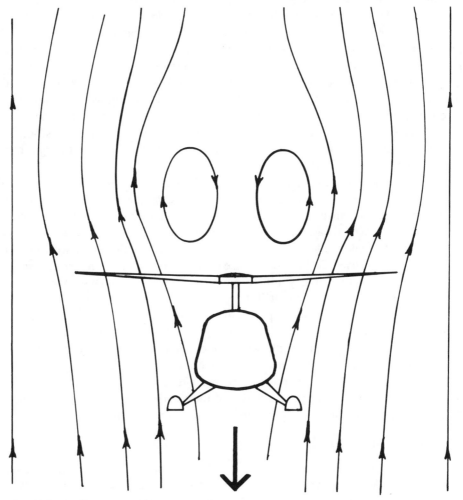

Fig. 8-2. *Airflow pattern in autorotation.*

Stay alert. Most helicopters will tend to drop the nose a little so you'll have to compensate by applying aft cyclic to prevent the airspeed from building up. In addition, when you lower the collective and chop the power, you reduce the torque and the helicopter yaws left; therefore, right pedal pressure is required to keep the nose straight.

The only other thing you have to watch out for is excessive rotor rpm. In some helicopters, at certain gross weights, at some speeds, and in particular atmospheric conditions, the rotor rpm might actually speed up too much. Memorize the rotor rpm limits for your particular helicopter and check the rotor tachometer frequently. Try to keep rotor rpm in the middle of the limits, so that you have room to error on either side. The solution to high rotor rpm is simple: Raise the collective slightly to increase the pitch on the blades. This will increase drag and slow the rotor down. Don't pull too much collective or you might cause an excessive decay in rpm.

If rotor rpm is on the low side after entering autorotation, check that you've pushed the collective all the way down. If it is down, that's all the rotor rpm you're going to get, so make the most of it. Either the helicopter has not been properly rigged or the weight and atmospheric conditions are conspiring to keep rotor rpm from building any higher; however, if rotor rpm is below the minimum allowed for autorotation, discontinue the maneuver by rolling on throttle ("power on") and return to base. Have the helicopter checked by a mechanic after landing because something is not right. Don't fly a helicopter that cannot provide the minimum required rotor rpm during autorotation.

After the descent has begun, it will feel like any other descent, except that the vertical speed indicator will be showing a much greater rate of descent than you've ever seen before, as much as 2,500 feet per minute in larger helicopters. The helicopter will fly just like it always flies, so just concentrate on holding the best autorotation airspeed and do some shallow turns. Before you know it, it will be time to level off and terminate the maneuver. Be sure to increase the throttle before raising the collective! If you raise the collective first, you'll find out why you must always lower collective when the engine fails: rotor rpm will decrease rapidly.

Roll on the throttle to match up the engine and rotor rpm needles again, and then raise the collective as you continue to add more throttle to keep rotor rpm in limits. Be careful not to add too much throttle when the collective is down or you could end up over speeding the engine, the transmission, and the rotor head.

I suggest you do your first power recovery with a constant airspeed to get a feel for the way the helicopter responds; however, the priority is to follow your instructor's directions. Level off, take a deep breath, and climb up to do it again (FIG. 8-3).

Flare-type Autorotations

After making at least one more successful autorotation like the first one, it's time to practice a flare-type autorotation. The beginning remains the same: collective down, throttle reduced, needles split, hold the airspeed, check rotor rpm. The power recovery is different.

Bell Helicopter Textron

Fig. 8-3. *Practice autorotations at altitude before trying them closer to the ground: Bell 206L LongRanger.*

Choose a barometric altitude that will simulate ground level, let's say 1,000 feet. The height above the ground that the flare should be started is going to vary under differing atmospheric conditions, at different gross weights, and from helicopter to helicopter, but let's use 50 feet for the sake of example.

At 1,050 feet, then, you'll start your flare. Move the cyclic smoothly aft. The nose will come up and the fuselage will act like a big wind brake, slowing the descent and the airspeed. At about the time the airspeed drops below translational lift airspeed is a good time to move the cyclic forward again to obtain a level attitude. As you come back to level, you should be at about 1,010 feet and the forward airspeed should be dropping below 20 knots.

In a real autorotation close to the ground, this is the time you would now pull the collective to use the inertia built up in the rotor system to slow the rate of descent even more and cushion the landing. Pulling the collective will increase the pitch angle on the blades and give you a few seconds of increased lift. Of course, this higher pitch angle also means more drag and without the benefit of an engine powering the transmission, rotor rpm is going to decay but fast. The good news is you'll be on the ground by then, if you did the flare right. Once you're on the ground, losing rotor rpm is immaterial.

During this practice autorotation, however, you don't want to end up with decreasing rotor rpm, you want to do a power recovery. So, just like the other power recovery you did, you'll bring the engine back up by rolling on the throttle until the engine and rotor rpm needles are once again matched. You do this while you're in the flare with the nose up and just before you level the machine. After you level the machine, increase the collective while adding throttle to maintain rotor rpm and Violà!, you're hovering at 1,000 feet out of ground effect.

There are a few things to remember about the flare.

First, as you pull the nose up, the increased volume of air meeting the rotor disc will cause rotor rpm to increase. Actually, this is good, because it will give you more energy you can use to cushion the landing. It can be bad if you allow rotor rpm to increase too much because damage could be done to the rotating components. If you're doing a real autorotation such damage will be the last thing on your mind. Nevertheless, when you're training, there's no reason to stress the machine unnecessarily. To keep the rotor rpm from spinning out of sight, simply nudge the collective up slightly. This has the added bonus of making your flare more effective. But be careful not to raise the collective too much or you'll lose an excessive amount of rotor rpm.

Tail rotor control is another thing to watch. When you entered the autorotation, recall that you had to use right pedal to counter the loss of torque. As you pull the collective up to effect a power recovery after leveling out, Newton's Third Law rears its ugly head once again and you'll have to use left pedal to counteract the torque of the main rotor. If you're doing a real autorotation, or a practice one all the way to the ground, you'll actually need more right pedal as you pull collective to cushion the landing because the loss of tail rotor rpm reduces its effectiveness.

CLOSER TO THE GROUND

Once you've mastered the technique of doing a power-recovery autorotation at 1,000 feet, then it's time to try one closer to the ground (FIG. 8-4).

More likely than not, you'll find it much easier because you'll have ground references to help determine when to flare and when to pull pitch.

There's one visual illusion you should be aware of, if you haven't noticed it by now; ideally your instructor has made you aware of it during his demonstrations or you have naturally noticed it. As you descend toward the ground from 1,000 feet or so, the ground doesn't appear to be coming toward you very quickly. Then, at about 100 feet, it suddenly starts to rush up at you faster than you expect. The faster your rate of descent, the more you'll notice this phenomenon and the harder it will be to judge your height above the ground.

The first few times you experience this when autorotating to the ground will probably be very disconcerting. Normal reactions are to flare too high in an attempt to stop the apparent rush toward the ground or to "freeze up" at the sight and flare too late. Just be aware of the fact that the ground will suddenly seem to rush up at you and rely on your altimeter to give you a more exact indication of your actual height. You'll get used to it after you see it happen a number of times.

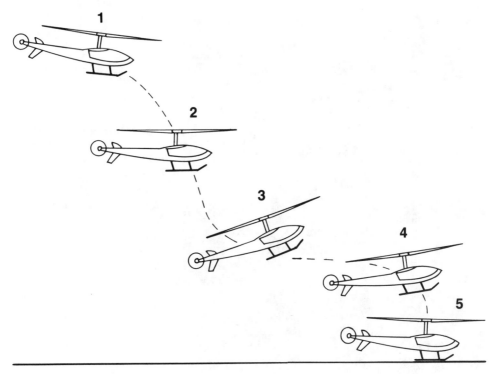

Fig. 8-4. *Flare autorotation and landing: (1) Lower the collective smoothly and decrease the throttle to cause the engine rpm needle to decrease below the rotor rpm needle. Add right pedal pressure to hold the heading. (2) Maintain the best autorotation airspeed by raising or lowering the nose with cyclic. (3) At the manufacturer's recommended height, use aft cyclic to flare to a nose-up attitude in order to decrease groundspeed. If doing a power recovery, roll on the throttle to match up the engine and rotor rpm needles. (4) After the desired groundspeed is reached, move the cyclic forward to level the aircraft. (5) Allow the helicopter to descend vertically and pull collective to cushion the landing, or stop in a hover, if doing a power recovery.*

Except for this visual illusion and the fact that you'll end up in ground effect instead of out of it, doing a power-recovery autorotation to a hover a few feet over the ground is just like doing one at 1000 feet above the ground (FIG. 8-5).

Practice autorotations to the ground should be done over a runway or to a large flat open area with a good surface, well clear of buildings and people, and with a good surface wind indication device. At a controlled field, be sure to request permission from the tower before starting autorotation training and keep a good look out for other traffic. Autorotations are nonstandard maneuvers and other pilots won't be expecting them. Follow your instructor's recommendations.

Fig. 8-5. *A practice flare autorotation to the ground is normally terminated in a hover by doing a power recovery. The Bell Helicopter Customer Training Academy in Fort Worth, Texas, teaches full touchdown autorotations: Bell 206 JetRanger.*

GOING ALL THE WAY

You can see that a power-recovery autorotation is just like a real autorotation except for the very last part: the touchdown. Even though the touchdown is a very important part of a real autorotation—and many people argue it's the most important part—the power-recovery autorotation is still a good training maneuver. But to practice the real thing, you have to go all the way down to the ground.

Full touchdown autorotations, as they are called, are not practiced by everyone. The U.S. Army and Air Force stopped doing them years ago when accident statistics showed that there were more injuries and damage from practicing autorotations than there were from actual autorotations. It simply made economic sense to stop practicing full touchdown autorotations and do power-recovery autorotations instead.

Touchdown autorotation advocates remain in the helicopter business. They agree that the first part of an autorotation is important, but they argue that the ball game is won or lost between the flare and the ground. Power-recovery autorotations don't train touchdowns. I think even the military agrees with this argument; they just can't justify all the bent training machines.

If you fly single-engine helicopters, autorotation practice is a must and touchdown autorotations are the best. If you fly twin-engine machines, touch-down autorotations aren't worth the risk. The chances of a dual engine failure are so small and the probability of a poorly executed touchdown auto is so high, that it just isn't cost-effective. Periodic training in power-recovery autorotations should be done, however.

Finding an instructor who will do full autorotations might be difficult. It is a potentially hazardous maneuver and few owners like to risk damage to a helicopter by letting inexperienced pilots do the maneuver. Their only insurance is a good instructor who is current and well-trained in touchdown autorotations. The most likely place to look for such instructors is at factory training schools and at large flight academies. The Robinson and Bell factory schools are two examples.

Before doing full touchdown autorotations, let's see how to do them from a hover.

HOVERING AUTOROTATIONS

Even if your instructor can't, won't, or isn't allowed to do full autorotations to a touchdown, he will be able to teach you hovering autorotations, which are done to the ground. These are really quite fun and so easy that the probability of damaging a machine is very slight (FIG. 8-6).

Fig. 8-6. *Hovering autorotations are not difficult and can be safely practiced with little risk of damaging the machine: Bell 206 JetRanger.*

What you try to simulate, of course, is an engine failure in a hover. This can be a critical time to lose an engine if you're in a high hover (*see* the dead man's curve discussion in this chapter), but from a low hover it's almost a sure thing. There's only one way you're going and that's down and from a low hover you don't have far to descend. To keep from smacking the landing gear on the ground too hard, you just need to pull collective to cushion the landing.

You want to keep the aircraft level with cyclic. Always try to land as level as possible, whatever kind of landing. The only other thing to worry about is the loss of torque when the engine dies. Because hovering is a high power maneuver and high power maneuvers require a left pedal input, you'll need to ease off the left pedal pressure and might need some right pedal to hold the heading.

The trick is to catch the movement right away and correct for it. Use whatever pedal necessary to keep the nose straight. You'll find after a while this comes naturally. If you ask a number of very experienced helicopter pilots which way the nose will yaw in a hovering autorotation, it would probably take them some time to figure it out; if it happened to them in flight, they wouldn't hesitate for a second to hold the nose straight with whatever pedal they needed, without ever thinking consciously about it.

All the Way Again

A full touchdown autorotation is just like a power-recovery autorotation right up to the point where you begin to roll on throttle in the flare. At this point, instead of rolling on the throttle as you pull collective, you check that the throttle is closed at the ground idle position to ensure that it doesn't inadvertently reengage the transmission and possibly cause an overspeed.

Now it's the energy in the rotor that provides the "power" to cushion the landing. Level the aircraft with cyclic and smoothly pull up the collective to further reduce the rate of descent and cushion the landing. Use the pedals to maintain heading. You'll probably need some right pedal, but don't think too much about it—just keep the nose straight by using your feet. From here on in, it's just like a hovering autorotation. When you're on the ground, hold the cyclic in the neutral position and lower the collective.

A touchdown autorotation is easy to write and talk about. It's hard to do. Until you get a good feel for it under your instructor's tutelage, you'll flare too high, then flare too low; you'll pull up the collective too soon, then pull it up too late; you'll level off too late, then level off too soon. You'll think you're doing everything just perfect—and it really will look perfect—and you'll land hard. It takes time and lots and lots of practice before you really start to feel confident. As you can see, it's very important to practice.

COMMON ERRORS

Common errors should be kept in mind:

- Flaring too high, like doing an autorotation from a high hover, will leave too much space between the helicopter and the ground. There might not be enough inertia in the rotor to prevent a hard landing and you'll be tempted to pull up the collective pitch lever too soon.

- Pulling up the collective pitch lever too soon will cause rotor rpm to decay while the helicopter is still too high above the ground resulting in loss of control and lift. Yaw control will be lost first, followed by cyclic control.

- Flaring too low will probably result in a touchdown speed that is too great and will leave little time for the application of up collective. It could also result in a nose-up attitude at touchdown and possibly a ground strike with the tail.

- Pulling up the collective pitch lever too late will result in a hard, fast touchdown, maybe even a bounce, after which the helicopter will become uncontrollable as rotor rpm decreases. Don't be too conservative with your collective pull when it comes time to cushion the landing. You don't want to "waste" rotor rpm by pulling the collective when you are too high, but use what you have to make a soft touchdown.

- Landing level is extremely important. Touching down in a nose-up attitude could cause the tail to hit the ground, possibly damaging the tail rotor and resulting in a severe yaw, or cause a pitch forward on the landing gear that could flip the helicopter over. Leveling off too soon or with a nose-down attitude will cause airspeed to build up again and result in a high groundspeed at touchdown. A severe nose-down attitude and a high groundspeed at ground contact will definitely cause a roll-over.

- Touching down with a yaw and more than a few knots forward speed could also cause the helicopter to roll over.

AUTOROTATIONS—180 AND 360 DEGREES

You could, of course, do 90-, 270-, and 720-degree autorotations ad infinitum, but 180s and 360s are the most common and will provide you with all the elements you need to do any kind of autorotation. Actually, once you have mastered straight-in autorotations, doing any kind of turning autorotation will be a piece of cake. There are a few things to think about while you are turning, but once you're lined up with the touchdown area, into the wind, everything is the same as a straight-in autorotation.

Getting yourself aligned into the wind will be your primary motivation for doing a turning autorotation. Another reason would be the lack of suitable landing areas anywhere else except the spot directly below you or close by.

Perhaps you're at 3,000 feet when the engine fails and the landing area is right below you. Instead of flying away from the spot and then back toward it to lose altitude (and risk landing short because you miscalculated the glide distance), it would be more prudent to spiral down above it so you have a higher probability of landing in the area.

The amount of altitude you lose in a turn will be dependent upon a lot of factors, including angle of bank, altitude, temperature, gross weight, and the type of helicopter you're flying. As a rule of thumb, you can usually figure on losing 500 feet for every 180-degree turn and, if you haven't achieved this rate after the first turn, you'll probably be able to adjust your angle of bank (within the limitations of your helicopter) so that you do achieve this rate on the next turn. Using a 500 fpm-descent for every 180 degrees just makes it easier to calculate how many turns you'll be making before reaching the ground.

Remember two important things about turning during autorotation:

- In a turn, rotor rpm will increase above the value obtained in a straight-in autorotation because of the g-forces generated in a steep turn. The steeper the turn, the greater the g-forces and the more rotor rpm will increase. If it starts to build too high, you can correct this by raising the collective slightly. Remember to lower the collective again when you roll out on final.

- Avoid the temptation to use pedal pressure to increase your rate of turn, unless absolutely necessary to align on final. Too much or not enough pedal will cause a skid or slip and both these conditions will increase the rate of descent and shorten the glide. Instead, use right pedal to keep the aircraft in trim.

To do a 180-degree autorotation, align the helicopter on a close downwind, 500 feet above the ground. Start your first ones at best glide speed, then work up to cruise speed as you get better. (Consult with your instructor for any variations.)

Abeam the landing spot, lower the collective, roll off the throttle, and start your turn. On entry, look out at the horizon, instead of at your spot. This will help make a coordinated turn. Go for an initial bank of 30 to 40 degrees. Check airspeed and rotor rpm and, if needed, make corrections to keep them in limits.

The combination of bank and autorotation attitude will make it seem like you're diving toward the ground more so than in a straight-in autorotation, so you might feel the temptation to ease back on the cyclic. Check airspeed first. If it's at the best glide airspeed, don't change the cyclic position.

Halfway into the turn, glance out at the landing spot to check your progress. Steepen the bank if you're too close; roll out a bit if you're too far away. Your goal is to roll out wings level on the centerline at least 50 feet or so above flare altitude.

Glance quickly at rotor rpm as you roll out and lower the collective all the way, if you raised it during the turn. From this point on it's a straight-in autorotation. Do a power recovery or full touchdown, whichever applies.

A 360-degree autorotation is done just like a 180, except you start out on final, 1,000 feet above your landing spot. A 720-degree autorotation is started on final at 2,000 feet, and so on and so on.

DEAD MAN'S CURVE

The ability to autorotate is a handy option to have up your sleeve, but, unfortunately there are some combinations of airspeed and altitude that do not provide enough time and altitude to do a decent autorotation.

Figure 8-7 is a *height-velocity* diagram, often called the dead man's curve. The manufacturer is required by the FAA to provide a height-velocity diagram for each helicopter type and include it in the pilot's operating handbook. The diagram in FIG. 8-7 does not apply to any particular helicopter and is for illustration purposes only.

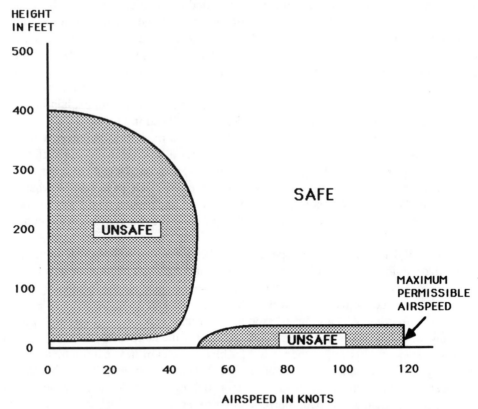

Fig. 8-7. *The height-velocity diagram reveals that if the engine fails while the helicopter is in the unsafe (shaded) area, it will probably not be possible to avoid crashing.*

The vertical axis of the diagram is in feet above the ground and the horizontal axis is airspeed in knots. The shaded area on the left side of the boundary line is the unsafe area. The unshaded area to the right of the line is the safe operating area. There's also a high speed segment of the diagram that is also shaded and therefore unsafe.

From any point on the boundary line, an average pilot should be able to manipulate the controls correctly, enter a normal autorotation, maintain rotor rpm, and land without damage to the helicopter or injury to himself (provided there's a halfway decent place to put the craft down). Inside the dead man's curve, the manufacturer's test pilots and lawyers don't expect the average pilot to be able to recover from a complete power failure without some damage to the helicopter. This is not to say that it's impossible to avoid bending the aircraft from every point inside the curve; however, it will take extraordinary skill and a lot of luck to do a noncrash autorotation from a height-velocity point inside the curve. From the boundary line outward (up and to the right) it generally gets easier to make an acceptable autorotation because there's more room for error.

An important point about height-velocity curves needs to be emphasized. The generic curve illustrated here and the one you'll find in the pilot's operating handbook are not valid for all conditions. Such factors as gross weight, temperature, altitude, and pressure will change the shape and size of the dead man's curve. Usually, the curves you find in flight manuals are drawn for average gross weights at sea level pressure and for a standard day (59 °F, 15 °C). Generally, as gross weight goes up and air density goes down, the height-velocity curve expands outwards, like a balloon. Either the graph itself or an explanatory section in the flight manual should explain how the graph changes under different conditions. Read it and heed it.

The worst place to be when the engine fails is in a medium-high hover with low forward airspeed and no wind. Recall from chapter 1 that there's a big difference between airspeed and groundspeed. You can have zero groundspeed, in other words be hovering over one spot, and have a high airspeed—your airspeed will be whatever the wind is. With 30 or 40 knots of wind in a medium high hover, you'd actually be in a fairly safe position if power were lost. The point on the diagram corresponding to 300 feet and 40 knots is the safe area.

But at 300 feet with zero airspeed, you'd be in the unsafe area. Medium-high, zero-airspeed hovers put you between the proverbial rock and hard place.

The rock is quickly decreasing rotor rpm. The hard place is the fact that lowering the collective is going give you a very high rate of descent.

Medium-high hovers are out of ground effect and therefore require more power than any other helicopter maneuver. This means that the collective pitch is pulled all the way up to your left armpit, the main rotor blades are at or near maximum available pitch angle, and the engine is giving all it's got. When the engine quits, the blades are taking an enormous bite out of the air. Think of sticking your hand out the window of a moving car and holding the palm perpendicular to the ground. Your hand is like a big wind brake.

The rotor blades aren't perpendicular to the ground, but they do have a very high pitch angle. This pitch angle causes them to slow down much quicker than if the helicopter was in forward flight. Consequently, rotor rpm decreases very fast. So fast that even if the pilot immediately recognizes the engine has failed and lowers the collective all the way down without delay, there still isn't enough time for the rotor rpm to build back up to a normal rpm before the ground comes rushing up to meet the bottom of the helicopter. If the pilot doesn't realize the engine has failed for a second or two and doesn't react right away, which is more probable, rotor rpm will decrease even faster and there's even less chance of a non-crash landing.

You don't want to descend faster than you have to, but you really don't have much choice. If you don't lower the collective right away, you'll very quickly lose rotor rpm and yaw control. If you lower the collective, you'll descend like a rock.

The best you can do is opt for the lesser of two evils. It's better to have some control over the helicopter and a high rate of descent, than to hope for a softer landing with no or little control. In other words, lower the collective, get back as much rotor rpm as possible, and make the best landing you can.

If you must crash, the best way to do it in a helicopter is coming straight down. Helicopters are designed to absorb a good deal of vertical loads. The landing gear struts or skids take the force first, then the fuselage and frame, then the seats. The machine might end up looking like a piece of modern art, but you'll probably be able to unstrap and walk away from the wreckage before collapsing from shock.

The point is to avoid the situation in the first place, which is what height-velocity curves are all about.

As you can see from the diagram, low hovers are in the safe area. If the engine fails, you descend, but not very far. There is enough inertia in the rotor system to allow you to increase the collective to add some lift and cushion the landing. Rotor rpm will decrease rapidly after you do that, but it won't matter because you'll be on the ground already.

To recover from an engine failure while in a high hover (a point above the height-velocity curve) is harder, but not impossible. Do two things: lower the collective and move the cyclic forward to lower the nose. The downward collective movement is to regain the rotor rpm, and from a high hover there's enough height to do this safely. The forward cyclic movement will cause airspeed to increase, which not only helps regain lost rotor rpm, but also makes the autorotation easier by reducing the rate of decent. Note the initial reaction of the helicopter to these two movements: a dive. As in recovering from a stall in an airplane, this maneuver requires some height above the ground or it can't work.

The other unsafe area on the height-velocity diagram is the high-speed segment. To fly faster and faster in a helicopter (above the best rate of climb speed), you need more and more power. This means the engine is working harder, you're pulling more and more collective pitch, and the blade pitch angle becomes greater and greater. At some point you pull more pitch and use more power than even in a high hover.

If the engine fails, rotor rpm will decrease rapidly because of the high pitch angle, just as it does in a hover; however, you do have an ace up your sleeve: airspeed. By reducing the collective and applying back cyclic, you can convert airspeed to altitude and regain and maintain rotor rpm during the rest of the maneuver.

"If one has this airspeed ace, why the high speed segment?" you might wonder. The reason is that low and fast doesn't give much room for error. If you are inattentive for a second too long or make only a small improper control input, there's a good chance you might make what accident reports call "an inadvertent descent into terrain" (FIG. 8-8).

Fig. 8-8. *Flying too low and fast can be just as dangerous as flying too high and slow. You don't have a large margin of error in case of engine failure: Phillipine Air Force McDonnell Douglas MD 500.*

Because the FAA Says So

Another very good reason to abide by the height-velocity diagram is the fact that the FAA has used it during enforcement actions against helicopter pilots to prove violations of Federal Aviation Regulations (FAR). The reasoning, which has been accepted by the National Transportation Safety Board, is that operation of a single-engine helicopter within the shaded areas of the applicable height-velocity diagram violates FAR 91.79 (Minimum Safe Altitude) and FAR 91.9 (Careless and Reckless Operation) because the pilot cannot safely autorotate in the event of an engine failure without undue hazard to people or property on the surface and that such operations endanger the life or property of another. Needless to say, violating FARs is a good way to lose your certificate.

Pay attention to the "dead man's curve." Manufacturers use the height-velocity diagram when devising the takeoff, landing, and other procedures for each helicopter model. If you always fly the helicopter according to the procedures specified in the pilot's operating handbook, you'll stay in the safe areas of the height-velocity diagram and keep yourself out of physical, and legal, trouble.

FINAL REMINDER

Remember this, if nothing else, about autorotations. You must maintain rotor rpm above the minimum limit or the helicopter will not fly. When the engine fails, LOWER THAT COLLECTIVE . . . .

9
Advanced Maneuvers

I often remember being asked in 1947 what I thought the helicopter could be used for. The answer, I thought, is like looking at a post office wall filled with hundreds of small boxes and finding in those boxes new ways to use this great machine. Perhaps we could open one a day, or a month, or a year. We've found many new uses, some good and some not so good, but we are still opening boxes to see what they hold.

Carl Brady,
former president and CEO,
ERA Helicopters,
Rotor & Wing International Magazine; 1991

THE VERSATILITY OF THE HELICOPTER IS ITS STRONGEST SUIT. HELICOPTERS can do so many things that it would be impossible to name them all. (I have tried, *see* FIG. 9-1) The things that will be done by helicopters in the future are only limited by man's imagination.

Fortunately, once you've learned to do a few advanced maneuvers, you'll be equipped to tackle many of the jobs on the list. This doesn't mean you'll be able to do every job. You won't, for example, be able to do aerial refueling, or long-line external lift work, or night instrument approaches to oil rigs, to name a few things. These all require special skills,

Aerial photography
Air-mobile operations—personnel transport
Air-to-air combat
Antisubmarine operations
Antitank warfare
Battlefield command and control
Battlefield observation and reconnaissance
Cattle herding
City-center passenger transport
Close air support of ground troops
Combat search and rescue
Corporate transport
Crop dusting
Crop seeding
Crop fertilizing
Delivery of oil dispersion chemicals on oil spills
Drug interdiction
Emergency relief operations
Environmental research
Erection of small bridges
Erection of transmission poles
Film stunts
Forest fire fighting
Geological surveys
Guarding coastal waters
Inter-hospital transportation
Lift loads to and from high buildings
Logging
Manned spacecraft launch and recovery support
Map surveying
Medical evacuations—air ambulance
Mine-sweeping
Move trees

News gathering
Oil pipeline inspection
Parachuting
Police work
Pollution inspection and control
Power line inspection
Power line insulator cleaning
Pull barges
Radar tracking
Radio and TV tower erection
Rescue from tall buildings
Retrieval of airborne drones and weather
 balloons
Search and rescue
Sightseeing
Special Forces operations
Supply lighthouses
Supply remote locations on land
SWAT team transport
Tow mine-countermeasures sled
Transport of checks and other banking materials
Transport drugs and organs
Transport of drilling rigs
Transport of personnel, supplies, and equip-
 ment to offshore oil platforms
Transport of harbor pilots to and from ships
Transport of sand to golf course sand traps
Transport Santa Claus
Tuna spotting
Unloading of ships
VIP transport

Fig. 9-1. *This list of things helicopters can do is long, but it's certainly not complete. Perhaps you can think of additional missions suited to the helicopter.*

additional training, and much experience, but you will have the basic advanced skills required for many helicopter operations.

This chapter covers many common advanced maneuvers: confined area operations, slope operations, maximum performance takeoffs, and pinnacle and ridge line operations.

CONFINED AREA OPERATIONS

A confined area is any area where the flight of the helicopter is limited in some direction by terrain or obstructions (FIG. 9-2). Examples are small clearings in wooded areas, parking lots, oil platforms, and sometimes even rooftop helipads (FIG. 9- 3). Confined areas can contain sloping ground or be on pinnacles or ridge lines. Basically, you should consider any off-airport or off-heliport landing area as a confined area until you ascertain it to be otherwise.

Fig. 9-2. *The helicopter's ability to operate from confined areas is one of it's important assets: Royal Air Force Boeing CH-47 Chinook.*

Fig. 9-3. *Offshore oil platforms, although man-made, have many elements of confined areas and pinnacles: Bell 212, Ekofisk Tank helideck, North Sea.*

Confined area operations require special techniques for both landing and takeoff. Before landing in any confined area for the first time, you should fly over the site at least twice and make one high reconnaissance and one low reconnaissance.

High and Low Reconnaissance

One of the first things the instructor will demonstrate and emphasize is how easy it is to miss seeing dangerous obstacles from a few hundred feet in the air. Wires and poles are particularly difficult to see. In some light conditions, they might be impossible to spot until you are almost directly upon them.

Always look out for the unusual, even when it looks like the usual. I was reminded of this fact on a flight I made in a Super Puma from Stavanger, Norway to Paris, France.

On a final approach to the downtown heliport at night I saw what I thought were edge lights outlining the helipad. I stopped in a hover over the lights, intending to descend directly onto the pad, but as I looked down I realized the lights were mounted on 3-foot poles, apparently some sort of visual approach guide and the paved helipad was actually behind me.

So, pay close attention whenever you make a landing at a new site, whether it's prepared or not.

The usual procedure at any site you're unfamiliar with is to fly at least one high reconnaissance (high reccon) and one low reconnaissance (low reccon). No rule says you should only fly one of each; fly as many as you like until you feel comfortable you've seen all you can see.

The high reccon is flown well above the obstacles, about 500 feet agl at a low cruise speed (FIG. 9-4). Check the size of the area, the condition of the surface, the degree of slope of the surface, the obstacles in and surrounding the area, and the wind direction. Try to get an idea where you'll run into turbulence.

You want to figure out the best way into and out of the site, taking into account wind and possible emergency landing areas if you have an engine failure on short final or after takeoff. Most of the time you should try to make use of the longest part of the confined area which means that sometimes the best approach and departure paths might not be into the wind.

After you do a high reccon, you'll be ready to do a low reccon. It's a good idea to get your prelanding checks out of the way before starting the low reccon. This way, if you must for some reason land immediately, you'll have everything done and won't have to worry about doing a checklist.

The low reccon is like a low approach with a go-around. Using the information you gathered during the high reccon, decide how you want to fly the approach and come in this direction on your low reccon. In effect, you'll be making a practice approach to the area while taking another look at the site from a lower height. After passing over the site in your high reccon, position yourself so that you are aligned on your proposed final approach course at 500 feet and about one-half mile from the site.

Fig. 9-4. *The high reccon is flown at a safe height above obstacles at a low cruise speed: McDonnell Douglas AH-64A Apache.*

Make a descent on final but don't descend below the height of the obstacles. Maintain cruise speed. If you've misinterpreted or underestimated the effect of the wind, you want to have airspeed, altitude, and power available to fly out of there.

During the low reccon, you should check for the same things you did during the high reccon. Choose the spot where you want to touch down and find a prominent object to help you identify it when you come around for landing again. Often you might have to modify your original plans after examining the area from a low reccon. Perhaps the touch-down site you picked out during the high reccon is too uneven or has too much slope. Maybe the wind changes direction as it curls around an outcropping of rock. Be flexible. If you need to change your initial plans, do so.

Most of the time it's not necessary to fly another low reccon even if you do decide to come in from another direction. If you prefer, fly another one. It won't cost more than a little time and fuel and will probably make your actual approach even better. You can always learn something else.

If you don't feel you need another low reccon, now's the time to land. Double check the prelanding checklist on downwind for the confined area.

Approach and Landing

The size of the confined area and the obstacles surrounding it will dictate the type and angle of approach to fly. Usually, it's a good idea to fly as close to a standard rectangular traffic pattern as possible in order to make everything as normal as possible. That way you won't be adding any new variables to a situation that is already different enough. If something happens, you will at least be able to start from a fairly normal condition.

Of course, often it's not possible to fly a standard pattern because of wind, weather, and obstacles. In such cases, you should fly the safest route possible under the conditions.

Plan to touch down in the lower quarter of the upper half of the area, if conditions permit, and adjust your approach angle accordingly. If it's windy, don't land too close to the obstacles on the upwind boundary of the pad because you might run into downdrafts. On the other hand, don't land too close to the obstacles on the downwind boundary because you'll have to make the approach too steep. Landing a little in front of the middle gives you a good compromise, and some room for error. Always pick one spot to aim for. This makes for a neater approach. It will also help you notice when you deviate from your intended approach path and angle, perhaps due to a change in the wind.

Be careful not too get too steep (FIG. 9-5). Confined area approaches are notorious for luring helicopter pilots into settling with power. (Recall settling with power in chapter 2.) Some confined areas seem almost made for catching helicopters. If the area is small, slopes downward on final, has high obstacles on all sides, is in the mountains, and has downdrafts, you'd better go in with a good power reserve or you could find yourself in a very precarious position.

As you descend, continue to inspect the area. Some training books call this the low reccon, but I think you'd be smart to make a separate low approach of the area, as described above. Of course, if you have landed in the area in the last day or two (any longer and new obstacles might be in place), you probably can get away with doing your low reccon on final approach. Stay alert and be prepared for sudden changes in the wind.

Unless you have to land in light snow or on a very dusty area, it's best to stop in a hover and closely inspect the ground beneath the helicopter. Does the surface look firm enough to support the landing gear? If the ground looks doubtful, lower the collective cautiously until you're sure that the helicopter won't sink too far. Pay particular attention to any rocks or stumps that might protrude enough to puncture the fuselage if the gear sinks into soft ground.

The worst situation you can get into is if one skid or wheel sinks and the other one doesn't. You absolutely don't want this to happen because the helicopter could easily roll over. If you feel one side sinking, lift the machine back up into a hover and find a better place to set down.

SLOPE OPERATIONS

Slopes are standard fare in unimproved, confined areas. Sloping terrain doesn't have to be very steep to be of concern to the helicopter pilot.

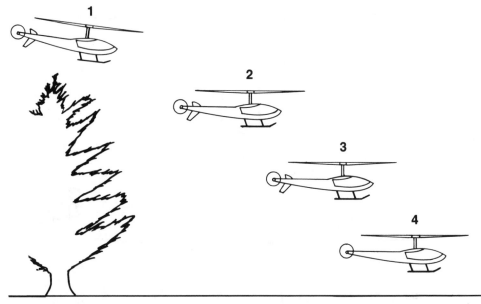

Fig. 9-5. *Confined area/steep approach (fly a high and low reccon for safety check): (1) Line up on final with normal approach airspeed, power, and rpm. (2) Lower the collective a greater amount than normal to start a steeper descent. Adjust the cyclic to maintain airspeed. Use the collective to maintain a constant sight picture (angle of descent). Pedals control the heading. (3) At the manufacturer's recommended altitude, apply aft cyclic to decrease groundspeed and start the landing flare. Increase the collective as the airspeed drops below translational lift so that the rate of descent does not increase. Do this flare high enough to avoid striking the ground with the tail. (4) Use forward cyclic and collective as needed to stop in a level hover over the landing spot.*

Despite their all-terrain label, helicopters really aren't very good on slopes. The FAA recommends that you do not land on a slope greater than 5 degrees. That's not very much. Believe me, a 5-degree slope will feel so uncomfortable that you won't want to attempt a steeper one.

Two hazards are associated with slopes. Hint: Both hazards are spinning at a high rpm. Never land the helicopter with the nose facing downhill because the tail rotor might strike the slope. Landing with the nose uphill is permissible, but now the main rotor gets too close to the ground. Uphill landings also make it more dangerous if you are loading or unloading passengers because the common procedure is to have people walk to the helicopter from the front.

Usually, the best way to land on a slope is cross-slope, like a skier traversing a hill (FIG. 9-6). Because the helicopter is more or less level in a hover, the upslope skid or wheel will touch the ground first. Skids are better than wheels for landing on a slope because their shape nestles in and holds the ground. A single wheel on the ground can easily act as a pivot point if you're a little sloppy on yaw or roll control.

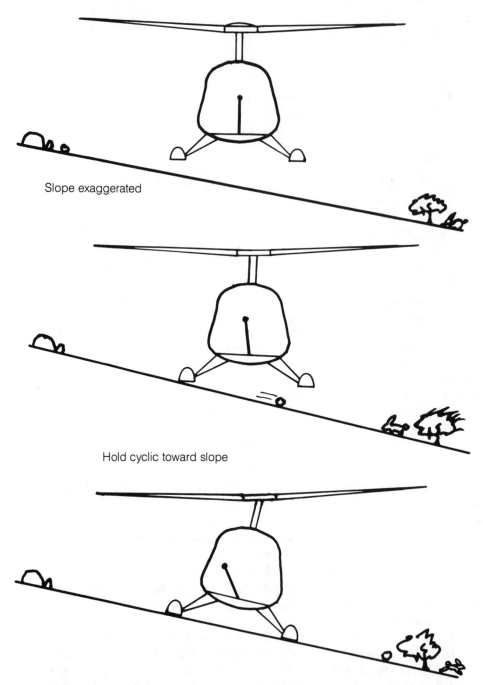

Slope exaggerated

Hold cyclic toward slope

Fig. 9-6. *Landing and taking off from a slope. The cyclic is held into the slope to keep the uphill skid or wheel firmly on the ground. Continue to move the cyclic toward the slope until the downhill landing gear is on the ground, too.*

Continue to lower the collective after the uphill skid touches, and hold your position steady with the cyclic and pedals because the fuselage will lean downhill and might cause the upslope skid to slide. Apply a small amount of cyclic toward the slope to counteract this tendency. Keep an eye on the clearance between the rotor tip and the ground on the uphill side as you come down. When the downhill skid touches the ground, continue down with the collective and add a bit more cyclic towards the slope. Be ready for an immediate takeoff if the helicopter starts to slide downhill.

A takeoff from a slope is the reverse of a landing. Center the cyclic and smoothly raise the collective. The downhill skid or wheel comes up first. Readjust the cyclic position as necessary to affect a level liftoff when the uphill skid comes off the ground. Remember to climb to an adequate height before turning the nose downhill so that the tail does not strike the ground.

Ground Reconnaissance

If the confined area is fairly large, you probably don't need to get out of the helicopter to do a ground reconnaissance. You only need to lift into a hover and air taxi to the takeoff position, watching where you're going. You can hover sideways or do a 180-degree turn and hover downwind, if the wind isn't too strong. It's usually best to hover sideways to the right if you're sitting in the right seat (sideways to the left if you're in the left seat) in order to have the best possible view of where you're going. If you think you'll need to hover backwards because of the wind or the nearness of obstacles, do a ground reccon.

If you leave everything running, be sure to engage the collective and cyclic control locks or increase the friction so the controls cannot be moved. It's easy to bump the controls as you climb out of the seat and if you accidently slip and really displace the cyclic, you could cause the main rotor to flap so low it hits the tailboom. Use good, firm hand holds and pay attention to feet movement.

Apropos to engaging control locks is the story about a United States Army student pilot who left everything running and forgot to lock the controls after landing in a remote, confined area on a ridge. After doing his ground reccon, he decided to visit Mother Nature. While he was in the bushes some distance away from the helicopter, the collective vibrated up enough to lift the little Hughes TH-55 off the ground. The helicopter did a respectable confined area takeoff before rolling on its side and crashing on the other side of the ridge.

As it happened, an instructor and student in another Army TH-55 were in the vicinity and saw the crash. They flew quickly to the site, landed in the same confined area, and got out to see if they could find any survivors. As they scrambled down the ridge, the first pilot emerged from the bushes and climbed into what he thought was his machine.

He took off, saw the crashed machine and the two pilots frantically waving on the hillside below, and radioed to operations to inform them of the crash. He then landed back in the same confined area, thinking himself quite the hero. It took the instructor and other student a good amount of arguing and a check of the maintenance log and tail number before they could convince the first student that he had taken their aircraft and his helicopter had crashed.

MAXIMUM PERFORMANCE TAKEOFFS

It came as a great surprise to me when I learned that helicopters do not normally take off and land vertically. I could hardly believe it. Isn't that what helicopters are supposed to do? Isn't that what they were invented for?

If you read aviation history, it's apparent that vertical flight was the primary dream and motivation of early rotary-wing designers and I suspect the discovery of translational lift must have come as some surprise to them. Here was their vertical-lift machine actually taking off better moving horizontally over the ground than going straight up into the air. They no doubt grasped the value of translational lift, but I'm sure they must have felt somewhat disappointed that a similar increase in lifting capacity didn't occur as one ascends vertically, too.

The lack of a "vertical translational lift" factor notwithstanding, vertical takeoffs and landings were still the main impetus for rotary-wing development. The idea of being able to operate from areas no bigger than the average driveway was enticing. The front cover of the February, 1951, issue of *Popular Mechanics* even shows a drawing of a suburbanite of the not-too-far-off future pushing his personal helicopter into the garage at home while another small commuter helicopter flies overhead. Things didn't work out the way the visionaries saw them.

A maximum performance takeoff is the closest thing you'll get to a vertical takeoff (FIG. 9-7). True, there will be times you'll have to climb straight up to clear the obstacles in your path, but you'll need all the power you can get to do it. Worse, you could end up placing yourself inside the dangerous part of the height/velocity curve and that's a definite no-no. You're going to have to evaluate conditions very carefully if you don't have at least some horizontal space to use to gain airspeed as you climb.

With a maximum performance takeoff, you use the vertical-lifting capability of the helicopter while still operating it safely. Because it requires more power than a normal takeoff, a maximum performance takeoff can rarely be done at maximum gross weight. Usually you'll have to figure on operating the helicopter at a lower weight if you plan to do a maximum performance takeoff, because you definitely want a good power reserve before doing one. Under extreme conditions, sufficient power to hover out of ground effect might be required to make this type of takeoff.

The angle of climb for a maximum performance takeoff will depend on conditions. Anything steeper than a normal takeoff path should be done using the maximum performance takeoff procedure. Remember that an engine failure at a low altitude and airspeed will place the helicopter in a position from which a successful autorotation might not be possible. In some cases, operation within the shaded area of the height/velocity diagram will be necessary during a portion of the takeoff path. The goal should be to get through this area as quickly as possible by increasing airspeed as soon as the obstacles are cleared.

Atmospheric conditions must be considered to determine power available. Wind will be the hardest to figure, and might ultimately be the most important factor. If at all possible, you'll want to take off into the wind, but it might be necessary to take off crosswind if that direction provides a better departure route. A downwind takeoff is asking for trouble and should be avoided unless there is absolutely no other way to get out of the area.

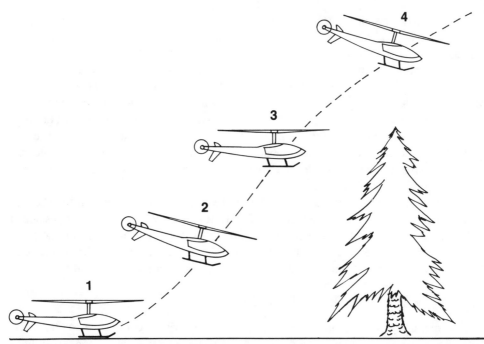

Fig. 9-7. *Maximum performance takeoff (do a hover check and land back on the ground): (1) Set takeoff rpm and increase the collective until the helicopter is light on the skids. Be ready for a left pedal input. (2) Increase the collective to maximum power, lifting vertically, then add slight forward cyclic to increase airspeed. (3) Climb at full power, adjusting the cyclic to control the angle of climb. (4) After clearing the obstacles, lower the nose and accelerate to normal climb airspeed. Continue a normal climb until reaching cruise altitude.*

Be particularly alert to the possibility of downdrafts on the lee side of obstacles. The stronger the wind, the stronger the downdrafts and associated turbulence. If the helicopter is flown into this area, climb performance will be diminished.

Before making a maximum performance takeoff, perform a hover check and pay particular attention to the power required to hover. This figure subtracted from the maximum power available gives you the power reserve. Your instructor should give you a rule-of-thumb minimum power reserve for your particular helicopter.

Lower the helicopter to the ground, being careful to keep the cyclic in the neutral position for hovering to eliminate the need for a large cyclic correction as you make the takeoff. Instead of departing from a hover, you're going to use all available power to leap the helicopter into the air.

On the ground, roll on throttle to set rotor rpm to the upper portion of the green arc and pull enough collective to get the helicopter light on the skids. Pause a second to clear the area above and in front of you and make one more check of the engine instruments. Now, pull collective to the maximum power setting and add full throttle up to the point where rotor rpm drops to the bottom of the normal operating range.

The helicopter should come off the ground vertically and because you want to gain airspeed as well as altitude, ease the cyclic forward slightly. Pedals, of course, control the heading and you should anticipate a good deal of left pedal input.

As you climb, adjust the angle of climb with cyclic inputs, aft cyclic to steepen the ascent, forward cyclic to make it more shallow. Check rotor rpm and if it decreases at all below the bottom of the green arc, nudge the collective down until it comes back up.

If you think the present angle won't clear the obstacles, don't try to "lift" the helicopter by pulling more collective if the engine is already at full power. All you will do is cause rotor rpm to decrease and when this happens, you'll lose lift and will climb even slower. Airspeed is what you need, so ease the nose over, gain some speed and then use this speed for better climb performance.

When you're clear of the obstacles, lower the nose by applying forward cyclic and accelerate in a level attitude to the best rate of climb airspeed. When you reach it, establish a climb attitude again and use full power to climb to a safe altitude above the obstacles, at least 500 feet above the ground. Once you attain 500 feet, you can reduce the collective to normal climb power and continue the climb or reduce it to cruise power and level off.

PINNACLES AND RIDGE OPERATIONS

A landing site on a pinnacle or ridge isn't necessarily a confined area; however, obstacles are common and must be dealt with. Pinnacle operations also apply to landings on pinnacle-like man-made structures, such as buildings and oil platforms (FIG. 9-8).

A pinnacle is any area from which the ground drops away steeply on all sides. A ridge line is a long area from which the ground drops away on one or two sides.

If obstacles are not present, the most hazardous parts of pinnacle and ridge line operations are dealing with the wind, the altitude of the site, and the rough and sloping terrain. If obstacles are present, then a combination of pinnacle and confined area operations will be required.

Rough and sloping terrain will dictate the need to hover, rather than making a landing to a spot; thus, you must be sure you have enough power to hover before making a pinnacle landing. High altitude and high temperature greatly reduce the performance of helicopters, so you must take these factors into account before landing at a site.

To be absolutely sure, use the "Power Required to Hover" chart from the aircraft flight manual. You get temperature from the OAT gauge and pressure altitude by setting the barometric altimeter to 29.92 inches (1013 millibars) and reading the result. (Do not forget to reset to current altimeter setting.) Make room for error, don't figure in a wind factor when determining power required to hover; if no wind, you have the correct figure; if windy (very likely), the increased performance will be gravy.

Always plan on making the approach straight into the wind, if possible; if not possible, try not to take more than a 45-degree crosswind, definitely not more than 90 degrees. If the only way into the site is downwind, don't go in unless you have a huge power reserve and there's an extremely important reason for landing at the site. Usually you won't have such a big power reserve, so no matter how important the reason is (even to

Fig. 9-8. *A roof-top heliport is a man-made pinnacle: Bell 222.*

save someone's life), you shouldn't attempt the approach. You won't help anyone and only make the situation worse if you crash trying to make the helicopter do something that can't be done.

When climbing up to or approaching any pinnacle or ridge line, do it at a 45-degree angle on the upwind side of the slope in order to take advantage of any updrafts (FIG. 9-9).

183

Bell Helicopter Textron

Fig. 9-9. *When climbing to a ridge line, make the approach path along the ridge line and into the wind, if possible. A steep approach might be necessary to avoid turbulence: Bell 222.*

Also, if you must break away from the climb, you'll need less power to descend away from the hill because of the additional lift from the wind.

The angle of approach will depend on wind effects at the site, from the shape of the pinnacle, and from any obstacles. Every situation is different. Sometimes a steeper-than-normal approach will work, other times a shallower-than-normal approach. I suggest trying a normal approach angle during the low reccon and see how that works. If you encounter excessive downdrafts, try to determine the cause; in some situations, a shallow approach will bring you under the worst turbulence and give the best angle; in other situations, a steep approach coming in over the downdraft area is better. It all depends on the site and the strength and direction of the wind.

Takeoffs from pinnacles and ridge lines are usually not as critical as landings. The best way to depart is into the wind and down the hill and such takeoffs are really quite fun, especially if the wind is stiff and the hill is steep. Lift off into a normal hover, do a standard hover check, then pull collective and go. With any wind at all you'll be through translational lift quickly and have plenty of power to climb; if you're going downhill already, you don't have to climb. As the ground falls away below you, simply hold forward cyclic pitch and let the airspeed accelerate. The higher airspeed will give you a faster departure from the area and give you a more favorable glide angle in the event you have to make a forced landing after takeoff.

If the wind isn't coming straight up the hill, opt for a downhill takeoff downwind or crosswind, as opposed to an uphill takeoff into the wind. Trying to climb uphill with low airspeed, even if the wind is helping you, will be a slow, ponderous maneuver. Instead, takeoff downhill and accept a greater nose-low attitude and the resulting higher rate of descent in order to build up airspeed. You'll have altitude on your side.

If the wind and obstacles are such that you must take off uphill and into the wind, accelerate and climb until you have at least 5 or 10 knots above translational lift and good clearance from any obstacles. Then turn downhill (and downwind), lower the nose, and accelerate to best climb speed.

ROOF-TOP HELIPORTS

A tall building is just a man-made pinnacle with very steep sides. Other buildings nearby and the presence of numerous people, many of whom might not be too excited about helicopters flying overhead, are factors that must be added into the equation.

I don't consider it particularly prudent to operate a single-engine helicopter over any city center area, although I know it is done all the time. The probability of a successful forced landing in a congested area is almost nonexistent and this makes operating a single-engine helicopter risky, not only for the pilots and passengers of the helicopter, but also for the people below. A crash in a downtown area by any helicopter does nothing positive for

Fig. 9-10. *Single-engine helicopters, like this Robinson R-22, can safely be operated to roof-top helipads in noncongested areas, but should not be used routinely in congested downtown areas.*

United Technologies Sikorsky Aircraft

Fig. 9-11. *Towing a mine countermeasures sled to secure a sea lane is an advanced maneuver that requires special training, but with the skills you have learned, you're well on your way to becoming a real pro: United States Navy MH-53E Sea Dragon.*

the industry as a whole. City governments react quickly to public outcry and have eliminated all roof-top helicopter operations in some cities after only one or two well-publicized accidents.

Only helicopters with at least two engines should be allowed into city center roof-top helipads; however, there are many other buildings with roof-top helipads that, because of their noncongested locations, are entirely suitable for single-engine helicopter operations (FIG. 9-10).

The difference between such a building and a remote area pinnacle (besides the fact it's man-made) is the presence of people; therefore, the inclination to dive off the structure in order to gain airspeed, which sometimes would be acceptable at a remote pinnacle, should be avoided when taking off from a building. Your goal should be to gain both altitude and airspeed, not just airspeed, because you want to get the helicopter up to a safe height as soon as possible. A safe height is higher over a populated area than over an unpopulated area. Reducing noise pollution is also very important over populated areas; higher altitude means less noise on the ground and better public relations.

Beyond these considerations, operations to and from roof-tops are the same as from natural pinnacles.

As stated earlier in this chapter, special helicopter applications require additional skills (FIG. 9-11), but if you can master the advanced maneuvers presented here, you will be able to do most operations that helicopters are called upon to do today, and be well on your way to flying like a pro, whether commercially or privately.

Speaking of being a real pro, being prepared for any eventuality is one characteristic of a professional pilot. The ability to troubleshoot and react to emergencies calmly and correctly is perhaps one of the most important skills full-time pilots must develop. Common helicopter emergencies are next.

10
Emergencies

The thing is, helicopters are different from airplanes. An airplane by its nature wants to fly, and if it is not interfered with too strongly by unusual events or a deliberately incompetent pilot, it will fly. A helicopter does not want to fly. It is maintained in the air by a variety of forces and controls working in opposition to each other, and if there is any disturbance in this delicate balance, the helicopter stops flying, immediately and disastrously. There is no such thing as a gliding helicopter.

This is why being a helicopter pilot is so different from being an airplane pilot, and why, in generality, airplane pilots are open, clear-eyed buoyant extroverts and helicopter pilots are brooders, introspective anticipators of trouble. They know if something bad has not happened it is about to.

Harry Reasoner
February 16, 1971

HARRY REASONER APPARENTLY NEVER HEARD ABOUT AUTOROTATIONS. On the other hand, I have to agree with him about helicopter pilots being "anticipators of trouble." I think all pilots should be that way. The "Basic Four-Step Aircraft Emergency Procedure" (FIG. 10-1) was introduced in chapter 8. It is time to discuss it in more detail.

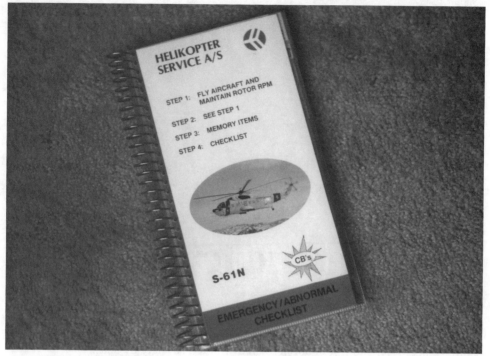

Fig. 10-1. *The Basic Four-Step Helicopter Emergency Procedure; front cover of Helikopter Service emergency checklist.*

EMERGENCY PROCEDURE EXPLAINED

To reiterate, with any emergency or unusual occurrence in any type of aircraft, your main concern is to keep the machine flying safely in the air until you can put it on the ground. Pay attention to heading, altitude, and airspeed—that's what "Fly the aircraft" means. In a helicopter you have one more thing to remember so you can take care of the above three parameters: "Maintain rotor rpm."

Step #1 is your life-saver (FIG. 10-2). So many well-documented accidents have been caused by pilots' failure to fly the aircraft first and take care of the emergency second that it is almost unbelievable. One in particular comes to mind. An airliner crashed into the Florida Everglades because the entire four-man flight crew was trying to trouble-shoot an unsafe nose-gear landing light indication. The autopilot was inadvertently disengaged and no one noticed the slow descent to the ground. The accident investigation board determined that the landing-gear warning was due to a burned out light bulb, but the airliner crashed because no one was flying the aircraft.

In a helicopter, maintaining rotor rpm is an integral part of flying the aircraft. The two are inseparable. Without rotor rpm, the helicopter cannot fly. If you are not flying the helicopter properly, rotor rpm is difficult, if not impossible, to maintain.

United Technologies Sikorsky Aircraft

Fig. 10-2. *Test pilots, like those flying this Sikorsky H-76, know their first concern during any emergency is to fly the aircraft.*

My first experience in how not to react to an emergency happened while I was a young lieutenant in the Air Force.

I was sitting in the jump seat of an HH-3E Jolly Green Giant helicopter when the right engine caught fire. A major was flying in the right seat although it was a captain in the left seat who was designated aircraft commander for the flight. (Usually the aircraft commander flew in the right seat.) On this occasion, the captain, who was chief training officer and an instructor pilot, was checking out the major in parachute operations. The major was not an instructor pilot. He had just arrived to the unit and was soon to be the unit operations officer. He would then be the captain's direct boss and, therefore, the one who would write his officer evaluation report. So this wasn't a normal pilot/copilot relationship. (Chapter 13 deals with similar problems in more detail.)

The fire warning light flashed on, the flight mechanic in the cabin said, "There's fire coming out of the right engine," and chaos broke lose in the cockpit. The captain saw the problem first, yelled "Engine fire," and grabbed the cyclic and collective. The major didn't want to release the controls and started spouting out portions of the emergency checklist. Each reached up to shut off the engine with the fire (the engine controls are on the overhead console in the HH-3E), but neither did. Hands flew all over the place. They both tried to adjust the collective to maintain rotor rpm, but when one lowered the collective, the other one raised it. The engine indications fluctuated so much that, from my position in the jump seat, I couldn't tell which engine was good and which one was bad.

After nearly a full minute of confusion during which each pilot tried to do everything himself, they finally agreed on which engine to shut down and which fire extinguisher to activate. Fortunately, they chose correctly.

Had one of them concentrated on flying the aircraft (the first step) and the other on doing the emergency procedure, the whole thing would have been done in a much more efficient and safe manner.

Etch that first step into your memory right now, so it is in your unconscious whenever you fly. If something happens, it should pop out and flash a distinct message to you: "Hey, shouldn't you be keeping an eye on altitude, airspeed, heading, and rotor rpm while you work on that warning light?"

I've seen hundreds of pilots react to thousands of emergencies during my time as a simulator instructor. You can troubleshoot and solve the most intricate hydraulic or electrical problem, but it will all be for naught if you allow the aircraft to descend into a mountain.

Step #2 requires no explanation. It's there to reemphasize the importance of the first step.

Step #3 refers to the memory (or must know) items in every emergency checklist. These items you must know by rote because they require immediate action. Memory items are usually the first two or three steps, and sometimes more, of the more important specific emergency procedures. Often the steps are printed in **boldface type** or are ┌ surrounded by a box ┐ to make them stand out (FIG. 10-3).

Every helicopter will have a few emergencies that need correction so quickly there isn't time to pull out and refer to an emergency checklist. A common problem with many checklists is too many of these must-know items. A well-constructed emergency checklist will keep the must-know items to a minimum because it's counterproductive if pilots are required to memorize too many things.

A few generic emergencies that have immediate action items and are common to all helicopters are: complete engine failure, engine fire, tail rotor failure, and hydraulic problems (if the helicopter has hydraulically-assisted controls). Other conditions require immediate action, too, such as low rotor rpm, settling with power, and ground resonance, but these are usually not addressed in emergency checklists because they should be avoided at all times.

Commit the must-know items to memory soon after you start flying a particular type or model of helicopter and test yourself on them often. Flying is not 100 percent excitement all the time. During the calmer periods, imagine an engine fire. What are

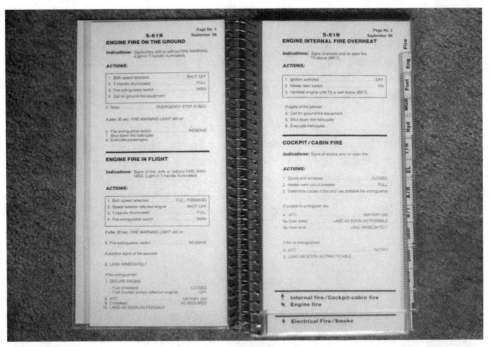

Fig. 10-3. *Memory items (for immediate action) in the Helikopter Service emergency checklist are surrounded by a box.*

the indications you can expect? What are the memory items for this emergency? Place your hand or finger lightly on the switch, handle, or control for each checklist item. This will instill in your mind an awareness of objects and actions instead of only words in the checklist. Do this with all the emergencies that have memory items until you can repeat the actions without hesitation. And test yourself periodically, say once every week, so you don't forget them.

Not all emergency procedures have immediate action items. This means there are a lot of things that can fail that you really don't have to get too excited about. Sure something's wrong and you have a problem and it might even require you to turn around and go back to base, but it's not so critical that you have to get all shook up and start switching things on and off.

One of the worst things you can do is start switching things on and off indiscriminately because the odds are you'll grab the incorrect switch, and if you don't make matters worse, you might inadvertently delay activating the remedy or deactivating the cause.

A Good Rule

This is a good time to introduce another general rule or principle that applies all the time in the cockpit.

193

Never switch anything on or off without first checking that your finger is on the right switch (FIG. 10-4).

This might sound very simplistic, but if you follow it religiously, you'll save yourself a lot of unnecessary trouble. Don't ever blindly reach up or reach over to switch something on or off. What you think is the heater might be the ground inverter or the landing light. You can put your finger there, but visually check it before you move the switch.

One memorable time in an S-61N simulator, I watched a senior captain with more than 10,000 flight hours very confidently reach up and shut down the good engine after the other engine had failed on takeoff. He realized his mistake as soon as he had the throttle back, but by that time, it was too late. Both engines were then shut down and the helicopter was flying at such a low altitude and airspeed the other pilot never even had a chance to lower the collective to enter autorotation. Had they been in a real

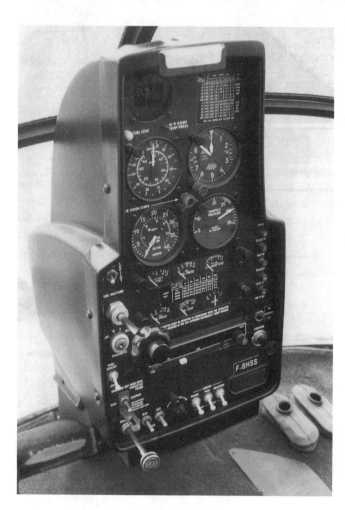

Fig. 10-4. *A good rule in every cockpit: Never switch anything on or off without first checking that your finger is on the correct switch: Schweizer 300 instrument console.*

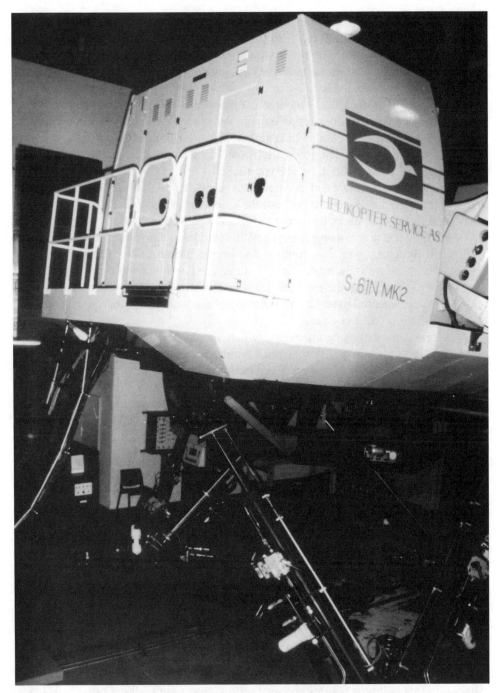

Fig. 10-5. *Simulator training is great for exposing unsafe habits that all of us acquire after many hours of uneventful flight operations: Helikopter Service S-61N simulator.*

aircraft, the crash would have been fatal. The captain could only look at the other pilot and say, with the utmost humility, "I'm sorry. I'm so sorry." Had he checked his hand before retarding the throttle, he never would have made the mistake.

Such incidents illustrate the value of simulator training (FIG. 10-5). I'm sure this captain relearned a valuable lesson that day and was much more aware of what he did with his hands in the cockpit. Some instructors even advocate immediately sitting on your hands when you first discover an emergency to keep those busy little fingers from doing something they shouldn't. That's okay if you have an autopilot or a two-pilot crew, but if you're by yourself you really should keep your hands on the controls. But the point is well taken, "Look before you flick."

Step #4 refers to the printed emergency checklist. After you do the memory items, if there are any, stop and pull out the emergency checklist. Don't do anything else—keep flying the helicopter—until you get the printed checklist.

Ideally, by this point in any emergency, the imminent danger has passed. The aircraft is still flying and you're in control. The immediate action items have been taken care of. Yes, there are a few more things that must be done, but you don't have to rush. Don't try to do more than the memory items, even if you think you know exactly what to do next. The potential to make a mistake is just too high. The consequence of doing the wrong thing is just too great.

Take out the emergency checklist, which should be stored in an easy-to-reach place in the cockpit, and verify that you have the correct checklist for that particular malfunction. It doesn't help matters if you perform the emergency items for the wrong failure. (It has been done.)

Most checklists show or tell what indications you can expect for the various failures. Confirm that you have those indications. An important word of warning: In the heat of the moment, it's very easy to misinterpret indications and see what you want to see, or expect to see. Be very critical of what you think you're seeing and double-check that the action is correct. Then, if there are any memory items, confirm that you did these properly. Only after these actions should you perform the rest of the checklist.

Look at each item and the action required. Find the switch or lever or handle or whatever it is you need to move and put your finger or hand on it. Confirm visually that you have the correct switch, lever, or handle. Then do what the checklist says to be done.

It might sound like a time-consuming process, but it really doesn't take much time. As the adage goes, "If you don't have time to do it right the first time, how will you find the time to do it over?"

TAIL ROTOR SYSTEM FAILURES

One of the most feared emergencies is a malfunctioning tail rotor system. Although some tail rotor failures are very serious, most are really not that bad. If you do the procedure correctly, you should be able to put the craft down with minimal damage and no injuries.

Tail rotor system malfunctions are normally divided into two categories:

- Control system failures include those malfunctions that reduce, partially or totally, the effectiveness of the pilot's control inputs to the tail rotor. The tail rotor is still rotating and will continue to rotate. These failures are less serious, although with improper pilot technique they could result in a crash.

- Drive system failures include such things as complete stoppage of the tail rotor, loss of a tail rotor blade, and separation of the tail rotor from the aircraft. Obviously, drive failures are much worse than control failures. The potential for a crash landing is much greater than with control-type tail rotor failures, but there still is the possibility of making a noncrash landing.

TAIL ROTOR CONTROL SYSTEM FAILURES

A tail rotor control failure could be caused by something as simple as a dropped writing pen, a forgotten screwdriver stuck in the pedal mechanism, control cable binding, or a jammed control in the pitch change mechanism. Theoretically, the pedals could stick in just about any position: left full forward, right full forward, both neutral, and every position in between. Chances are you won't notice the pedals are stuck until you try to move them out of the current position. For example, if they get stuck in cruise, you probably won't discover the problem until power is reduced for descent.

The collective pitch setting is the main determinant of tail rotor pedal position. The more power you are pulling, the more you need to counteract torque, and the more left pedal you need; less power, more right pedal. The extreme on the one hand would be a full-power takeoff or an out-of-ground-effect hover: full left pedal. The other extreme would be a steep descent in a lightly loaded helicopter in a right bank: a good deal of right pedal, perhaps even full right pedal.

The position that the pedals become stuck in will determine your response to the situation. A stuck left pedal will require a different action than a stuck right pedal. Let's see what action is required in both situations.

Let's say the left pedal becomes locked while you are in cruise, which would place it forward of neutral. As you lower the collective to start a descent, you notice you can't compensate for the reduced torque effect by easing off the left pedal and the helicopter consequently yaws to the left. You determine that the pedals are locked. Because the power used in cruise is usually close to power required to hover, the main problem you have now is the descent. Once you get close to the ground and pull in hover power, you'll need nearly the same amount of left pedal that is now locked in and you'll be able to hover with little or no yawing (FIG. 10-6).

You can cause the helicopter to descend three ways. First, increase airspeed while maintaining the same power setting. Simply ease the cyclic forward to lower the nose. As you get down to normal approach altitude, carefully bring back the cyclic to reduce the airspeed without climbing.

United Technologies Sikorsky Aircraft

Fig. 10-6. *If the Sikorsky SH-60B SeaHawk experiences a tail rotor control failure, the position of the tail rotor pedals will determine if the pilots can make a safe landing on the helideck of the* U.S.S. Crommelin *in the background.*

The second way is to reduce power and engine rpm and accept the resulting left yaw. As long as you maintain at least 50 to 60 knots, the streamlining effect of the fuselage will prevent the aircraft from yawing excessively and you'll be able to make normal turns.

The third way is to reduce rotor rpm slightly below the normal operating range, but not reduced outside caution range, and lower the collective a corresponding amount to cause a descent. This works because the main rotor produces less lift at a lower rpm and therefore creates less torque effect. This method is very effective, but must be done very cautiously. It should not be attempted without the benefit of instruction.

Regardless of the descent method, maintain airspeed above normal approach speed until approximately one-third down final approach. At this point, start a slow deceleration, using as few collective changes as possible, so that you arrive in a low hover above your expected touchdown spot just as translational lift is lost. Pull collective to stop the rate of descent and allow the helicopter to make a landing to a spot. Don't try to hover even though you might be able to, because descending from the hover could create a problem. Get the helicopter on the ground as soon as possible.

Because reduced-power maneuvers, such a descents, require a right pedal input, you'll usually be descending if the right pedal becomes locked forward of the neutral position. Hovering is therefore impossible because the higher power required to hover will cause the helicopter to yaw left, usually at a rapid rate. The way to get the machine down with a stuck right pedal is to do a running landing, which requires little or no increase in power.

Level flight and climbs are a problem with a stuck right pedal because you cannot counteract torque due to increases in power by adding left pedal. There are ways to do it, though.

If the pedal locks while you are descending at a high speed, say 100 knots, you probably could obtain level flight by reducing the airspeed to a low cruise speed, for example 70 knots, as long as the rate of descent was not too high to begin with. To obtain the best possible climb, reduce airspeed to the best-rate-of-climb airspeed.

If you don't climb or can't even maintain a level altitude at the best-rate-of-climb speed, the only thing left to do is to increase collective. Do this carefully and in small increments because the helicopter will yaw right. The more collective you pull, the greater the right yaw. If you really must gain some altitude, hold your airspeed at the best-rate-of-climb speed, pull collective, and allow the helicopter to spiral upwards in a right turn.

Because you want to do a running landing, you'll have to find an airport or some place with a long flat area. Do a shallow to normal approach and hold airspeed as close to best-rate-of-climb speed as the locked pedal position will allow (in other words with as little yaw as possible). This way, if you need to make a go-around, you'll already be at the best possible airspeed.

Experiment with small collective inputs. You'll find that a slight increase in collective will cause the nose to move left, and a slight decrease will cause the nose to move right. Now, freeze the collective and tell yourself you are not going to move it again until just before touchdown if necessary to make a small yaw input.

Aim for the middle of the first third of the runway or landing area, using cyclic control for alignment. At approximately 50-100 feet, depending on aircraft type, begin bleeding off airspeed so that you arrive a few feet over the ground just above translational lift. Whatever you do, don't allow your airspeed to decrease below this amount until you are on the ground.

Touchdown one of two ways, depending on your engine controls. If your helicopter has a motorcycle-type throttle grip, roll off some power and the helicopter will settle to the runway like a floating leaf. Rolling off the throttle will usually kill any left yaw and might even give you a slight right yaw. If it does, increase collective very slightly. This will eliminate the right yaw and cushion the landing. After ground contact, carefully reduce collective and throttle as necessary to maintain alignment with the runway and reduce forward speed.

If the helicopter has an overhead throttle, or fuel flow control lever, as many turbine helicopters do, reducing the throttle is impossible if you're the only pilot. For a two-pilot crew, it's a difficult exercise in crew coordination. The landing then becomes a very precise balancing act between cyclic and collective inputs. The key is to find an airspeed that provides a 300-400 fpm descent and no yaw and to hold this airspeed all the way to the ground. The only kicker is ground effect, which will cause the helicopter to level out a few feet above the ground. Now the balancing act begins.

You must bleed off a few knots of airspeed to get down. If you bleed off too much airspeed too fast, the helicopter will start to descend too quickly toward the runway. If you try to correct by increasing collective, you'll induce a big left yaw, airspeed will drop even faster, and you'll pull more collective. Soon, the situation is hopeless.

Very, very carefully apply aft cyclic pressure as needed to bleed off just a few knots at a time. If you do it correctly, you'll touchdown just at effective translational lift airspeed. The aircraft will still want to yaw toward the right as you lower collective, but when you're on the ground, the friction between the skids or wheels and the ground will help reduce this tendency.

Precise control inputs are needed to make decent landings with control-type tail rotor problems, but such landings are possible. As I mentioned before, tail rotor control failures are not that serious, but with improper pilot technique they can easily cause a crash.

TAIL ROTOR DRIVE SYSTEM FAILURES

Complete stoppage of the tail rotor, loss of a tail rotor blade, and separation of the tail rotor from the aircraft are extremely serious emergencies (FIG. 10-7), so serious that they are impossible to effectively simulate in a real aircraft. The best an instructor can do in training is push the right pedal cautiously forward to the stop, say "Simulated tail rotor failure," and have the student do an autorotation to a power recovery.

Fig. 10-7. *Damage to a tail rotor blade might cause vibrations that are so severe that the entire tail section separates from the aircraft: S-61 tail rotor.*

With the development of good, six-axis, visual helicopter flight simulators, more thorough training in tail rotor drive system failures is possible. It's still not the same as the real thing, but there isn't a pilot around who would willingly try the real thing. It's scary enough in a simulator.

What a simulator shows us is that a tail rotor drive system failure is controllable to some extent, if proper actions are taken quickly enough. The proper action, and the only proper action, is to enter autorotation immediately.

This is fine if you are in cruise, but if you are in a hover or just taking off or landing, it's almost impossible to land without crashing. Even from cruise flight, the probability of crashing is very high because the autorotation will be unlike any normal engine-out autorotation you've ever done. But even though you might total the aircraft, you'll probably be able to land with a slow forward speed and rate of descent that minimizes the consequences.

Fortunately, most tail rotor drive system failures do not come unannounced. Many helicopters have chip detectors installed in the intermediate and tail gearboxes (FIG. 10-8). Flakes, fuzz, and other metallic particles are attracted to a magnetic plug in the oil sump of the gearbox. When enough chips accumulate to make an electrical connec-

Fig. 10-8. *Tail rotor gearbox of an Enstrom F-28F. The nut-like structure connected to the electrical wire under the gearbox is the magnetic chip detector plug. The nut with the circular glass window (just above the chip detector plug) is used to check the level of the oil in the gearbox.*

tion, a chip warning light illuminates in the cockpit. The normal emergency action is to land immediately, following the emergency procedure checklist, if applicable.

Other indications are unusual noises and vibrations. Because it is spinning faster than the main rotor, a problem in the tail rotor will be felt as a high frequency vibration, often in the pedals themselves. Whenever strange vibrations are felt, an immediate landing should be made and the source of the vibrations investigated. It doesn't take long for a tail rotor to literally shake itself to pieces if it becomes unbalanced.

The ultimate indication of a complete tail rotor drive system failure is a sudden, hard yaw to the right. If this happens, there are two basic rules to remember.

First, enter autorotation immediately. This means lower the collective all the way down without hesitation. If you wait too long, the helicopter will quickly become uncontrollable. It might exceed safe angles of yaw and roll or rotate so much that it will be in rearward flight at a high airspeed. Recovery will be impossible.

Second, maintain forward airspeed in order to make use of the streamlining tendency of the fuselage. Aim for the best glide airspeed.

These procedures are easy to write and remember, but not so easy to perform. The autorotation should stabilize the right yaw, but probably won't eliminate it completely; to fly straight ahead, you'll need a left bank to compensate for the yaw; if you level the wings, you'll turn right; turning left will require a very steep left bank and might be impossible. Because of the bank, the rate of descent will be much higher than a normal autorotation. You have control of the aircraft, but just barely.

Once you've established autorotation and have some semblance of control, some flight manuals suggest that you carefully increase collective to see what happens. Perhaps you've only experienced a control system failure and can maintain level flight. (Some control system failures can seem like drive system failures.) If the helicopter begins to yaw right again, forget trying to regain level flight, lower the collective, and shut down the engine. You're going to have to make an autorotation all the way to the ground.

With a left bank to counteract the right yaw and a high rate of descent, the final phase of the autorotation is going to be tough. You won't have much time to choose a landing spot, but try to reach the biggest, flattest area you can. Plan to flare about 50 percent higher than the normal flare height. As you flare, roll the wings level, then do the rest like any other autorotation. Try to touch down with as little forward speed as possible because chances are, you'll hit with some side force and this will cause the helicopter to roll over. In any helicopter crash, you have a better chance surviving a straight vertical drop without horizontal movement.

MAIN GEARBOX MALFUNCTIONS

Ranking close to tail rotor drive system failures in severity are main gearbox failures. The reason is easy to fathom; if the main gearbox fails in some way, the rotor stops turning; if the rotor stops turning, the helicopter can't fly.

Because it is one of the components in a helicopter that can't be backed up by another system, main gearboxes are subject to exacting standards, constant inspection (FIG. 10-9), and numerous monitoring systems. Even in the smallest of helicopters, the pilot is provided with transmission oil pressure and an oil temperature gauge. Many helicopters also have magnetic chip detectors in the main gearbox, with exactly the same function as the chip detectors in the intermediate and tail rotor gearboxes.

Fig. 10-9. *Nonredundant components are subject to continual, periodic inspections. British Caledonian Helicopters' mechanics are inspecting the main gearbox and rotor systems between flights.*

Sophisticated helicopters have additional warning systems. The Aerospatiale Super Puma, for example, has two separate main gearbox low oil pressure warning lights, an emergency oil pump, a high oil temperature warning light, and a main gearbox fire warning system. Newer helicopters, like the EH-101, will come equipped with vibration monitors on the main gearbox (as well as other systems). These *health and usage monitoring systems* (HUMS) can detect subtle changes in the numerous vibrations emanating from the gearbox so that very small cracks can be discovered long before they turn into serious problems.

Like tail rotor problems, main gearbox malfunctions usually don't come unannounced. Abnormal temperature or pressure indications and unusual noises or vibrations will be your most likely guide that something has failed or is about to. Don't disregard these indications. Don't just hope they'll go away. Take them seriously and follow the emergency procedures for your particular helicopter. After you land, have the aircraft checked by a mechanic before flying the machine again.

Perhaps the most common main gearbox failure is a malfunction of the lubricating system. Transmissions need oil and the surest way to make gears seize up is to run them without oil. Although the newest military helicopters are constructed to run for a minimum of 30 minutes after a complete loss of main gearbox oil, don't count on a civilian machine flying that long. If you lose transmission oil pressure or see oil leaking out all over the place, land immediately.

ENGINE MALFUNCTIONS

If the engine fails completely, there's only one thing you can do: autorotate. (Autorotations are covered in chapter 8.) An engine can malfunction and performance can deteriorate without failing completely, however.

If the engine is running so roughly that the engine and rotor rpm needles are split, don't troubleshoot it, make an immediate autorotative landing to the most suitable area. After entering autorotation, do not attempt to return to powered flight and be ready for the engine to quit completely at any time.

If the engine is running roughly or loses power, but the engine and rotor rpm needles are not split, you should check a few things before making a precautionary landing.

The first suspect is fuel. "Am I running out of fuel?" is your immediate question. If the helicopter has several tanks, is one tank running out? If so, switch to another tank. Perhaps you've just refueled and have reason to suspect the fuel was contaminated or of the wrong type. Switch to a tank you know has good fuel in it, or make an immediate precautionary landing. Also be sure to check the mixture control. Perhaps you forgot to lock it in the full rich position and it has vibrated to a leaner setting.

Second, check the magnetos. Is the switch in the both position? If not, switch to both. Perhaps one of the magneto systems is firing out of sequence. You can check this by switching to right and left mags to see if there is any change in the operation of the engine. Be very cautious when you do this because the engine might fail completely when you switch to the faulty system.

If nothing helps and the engine continues to run roughly, proceed to the nearest available landing area at an altitude that permits a safe autorotation. The engine might fail completely at any time.

A frozen or stuck throttle in flight is an emergency condition that must be carefully evaluated by the pilot. The throttle might freeze under any power setting from full on to idle. A stuck throttle will require some experimentation to determine if a descent can be made without encountering a low rpm condition or an overspeed.

If a descent can be made without creating an rpm problem, make a running landing at the nearest airport or heliport. As soon as you are safely on the ground, shut down the engine.

If the throttle is stuck in such a high setting that a descent cannot be made without causing an overspeed, an autorotation must be done. Fly to the nearest airport or heliport and inform the controlling agency of your intention to make an autorotation. Line up on final approach into the wind and when you are sure you can make the runway, move the mixture control to full lean. Quickly return your hand to the collective and lower it immediately to enter autorotation. It will probably take a few seconds for the engine to quit, but don't wait for it. Get that collective down right away.

All helicopters have instruments to monitor engine conditions. At the very least, you'll find an engine oil pressure gauge, an engine oil temperature gauge, and a tachometer. Warning lights might be associated with these gauges. For example, there might be a low oil pressure warning light and a high oil temperature warning light. Gauge and warning light indications should be considered when troubleshooting a problem. If the gauge and warning light agree, for example the oil pressure gauge shows low pressure and the low oil pressure light comes on, you probably have low oil pressure; however, if the gauge and the warning light do not agree, there's a good chance that you have an indication failure.

As a general rule, if engine oil pressure drops below the minimum required, make an immediate precautionary landing and be ready for a complete engine failure on the way down. If any other engine instrument goes above redline, make a precautionary landing to the nearest available area and closely monitor the other engine instruments for signs of trouble.

A tachometer failure can fool you. In flight, the engine rpm needle and rotor rpm needle are matched. Because rotor rpm is so important, you'll be checking the tachometer frequently and will be alert for any deviation; however, if one of the needles should fail, there's no reason to get alarmed. Either needle by itself can provide the information necessary to safely continue flight to the nearest airport or heliport; therefore, you should not enter autorotation if either rpm needle suddenly goes to zero, unless the engine has obviously failed.

FIRES

A fire in flight is one of the worst things that can happen in any aircraft. Unlike surface vehicles that can stop quickly and evacuate all occupants in case of fire, aircraft have to get back down to earth before this can be done.

Engine Fire

Of course, there's always a big, hot fire going on inside the engine when it's operating properly. But when the fire gets outside the engine, it becomes a real concern for the pilot (FIG. 10-10).

Broken fuel lines, broken oil lines, and casing burn-throughs are the most common causes of engine fires. All helicopters have some kind of fire detection system on the engine. These are thermal devices that illuminate a warning light in the cockpit if

Fig. 10-10. *Modern turbine engines are built to withstand internal temperatures in excess of 1,000°C, but when things get hot outside the engine, the pilots must take action: Boeing 234 engine.*

they detect an unsafe high temperature. Although many fire warning systems are subject to false warnings, it's always prudent to take any fire warning seriously.

Because fire warning systems differ considerably from helicopter to helicopter, refer to the flight manual for specific procedures. This is definitely one procedure you should commit to memory.

All helicopters have an engine compartment fire extinguishing system consisting of a container of extinguishing agent (bromotrifluoromethane or freon gas) under pressure and the necessary plumbing to discharge the agent around the engine compartment. The gas extinguishes the fire by displacing the oxygen, but doesn't linger very long in the engine compartment. To maximize the effectiveness of the extinguishing agent, the flight manual might put an airspeed limit on use of the fire extinguisher. Memorize this limit and abide by it or the gas might not stay around long enough to put the fire out.

One very important thing to remember with any helicopter: If the emergency procedure instructs you to shut the engine down, first lower the collective and enter autorotation before you shut down the engine (if you're flying a single-engine machine). By entering autorotation first, you avoid the problem of having to reestablish normal rotor rpm; if you shut the engine down first and then autorotate, you might become distracted while dealing with the engine fire and forget to lower the collective as you shut the engine down.

Remember: Fly the aircraft and maintain rotor rpm, even with a fire.

Electrical Fire

Most of the time, electrical fires are confined to one component, usually a short circuit inside that component or an electrical connector. The standard flight manual procedure is to isolate the damaged component from the rest of the circuit. You can normally do this by turning the unit off or pulling its respective circuit breaker, or both to be extra sure.

Always check circuit breakers whenever you suspect any kind of electrical problem, and even when you don't. You'll be amazed how many things in an aircraft are hooked up to a circuit breaker.

The rule of thumb with a popped circuit breaker is to reset it one time and see what happens. If it stays in, fine, go ahead and use the component; a momentary overvoltage in the circuit probably caused the breaker to trip. If the breaker won't stay in or pops out again, don't use the component because something is wrong and you might aggravate the situation if you keep trying to reset the breaker or hold it in against its will. Leave the circuit breaker out and switch off the component for an added safety measure.

Another Good Rule

If something is broken, turn it off. For example, if the landing light burns out, turn off the landing light switch; if a fuel pump fails, turn off the fuel pump switch. In the vast majority of cases, it probably won't make any difference, but you never know. The open electrical circuit to the component could create more problems later on.

If switching off a burning component and pulling its appropriate circuit breaker doesn't stop the electrical fire, you'll have to eliminate power to the electrical bus that provides current to the unit. You might have to pull more circuit breakers or even switch off one or more electrical producers, such as the generator, transformer rectifier, inverter, or battery. In the very worst case, you might have to shut off all electrical producers. Some flight manuals even recommend switching everything off first in order to quickly stop the fire, then, after it goes out, you may turn various components back on one at a time.

A complete electrical shutdown is not a major problem in daylight. The helicopter will still fly without electricity, although all instruments requiring electrical power will not work. On the other hand, the nonelectrical instruments, such as the barometric altimeter, airspeed indicator, vertical speed indicator (all of which are pitot-static instruments) and the magnetic compass, will all continue to work (FIG. 10-11).

At night, switching off all electrical power is more interesting. Obviously, the cockpit becomes very dark. Always carry an alternate light source, ideally two flashlights with spare batteries, when you fly at night.

Your first indication of an electrical fire might be the smell. Although acrid and unpleasant, the smell of burning electrical wires is not hard to miss. Smoke might also be present.

Bell Helicopter Textron

Fig. 10-11. *A helicopter will usually keep flying even with all electrical systems switched off: Bell 206 JetRanger.*

If there's a lot of smoke, you'll probably have to ventilate the cockpit. Do this cautiously because the addition of fresh air might feed the fire with oxygen. Ideally, isolate the source of the fire before opening any windows.

Another indication of an electrical fire could be a plethora of unrelated warning lights and malfunctions. Wires from several different systems are often bundled together to make routing of the wires easier. If one or more of the wires becomes chafed and then short-circuits against another wire or the fuselage, the resulting build-up of heat could be enough to melt or burn the insulation around the other wires in the bundle.

The following incident, which occurred in a Super Puma, is typical of an electrical fire. The first warnings the pilots saw were the engine malfunction warning lights. As they worked on these, the autopilot cut out. Then warning flags appeared on a number of the navigation instruments. Finally, both generators dropped off line. What appeared to be multiple unconnected problems ended up being caused by a short-circuit of some of the wires in a bundle. Fortunately by cutting out all electrical power, they were able to stop the fire and eventually make a safe landing.

The bottom line: Treat electrical fires with respect and carry two flashlights at night.

Cabin and Baggage Compartment Fire

Unfortunately, helicopters are not well-equipped to fight fires in the cabin and baggage compartments. Even the largest helicopters carry only one or two small fire extinguishers. A few helicopters have fire detection systems in the cabin and baggage areas, but automatic fire extinguishing systems for these areas are very rarely provided. The common procedure in large helicopters is to send the copilot back to the cabin or baggage compartment to try to fight the fire.

In a small helicopter with one pilot, it's very difficult to fight a cabin fire and fly at the same time. If you have passengers, you might be able to instruct them. Basically, you do the best you can to extinguish the fire. Your main objective should be to get the helicopter on the ground as quickly as possible, shut it down, and evacuate.

We have reviewed the most serious helicopter emergencies. If you understand these, have a good working knowledge of every possible emergency that's explained in your aircraft's flight manual or pilot's operating handbook, and follow the four-step helicopter emergency procedure, you will be well prepared for any eventuality.

Although it's possible to drive a car without knowing what a carburetor is and it's possible to fly a helicopter without knowing how to put one together, the serious pilot will always strive to learn more about his machine. Knowing the parts and systems that make up your aircraft, how they work, how they interact with other parts and systems, and how they can fail will help you cope with the numerous minor glitches that occur and make you better prepared if something extremely serious happens.

Studying aircraft systems is an on-going task that starts in chapter 11.

11
Aircraft Systems

Aviation in itself is not inherently dangerous. But to an even greater degree than the sea, it is terribly unforgiving of any carelessness, incapacity or neglect.

Origin unknown

JUST AS THE FORDS, HONDAS, AND VOLKSWAGENS ARE CONSTRUCTED differently, so are Bells, Enstroms, and Robinsons designed and built differently. No two helicopter types are alike, even those built by the same manufacturer. This is why the study of aircraft systems is so important.

Not that you won't see general similarities in all helicopters. You will. And you'll find even more similarities in the engineering designs of each manufacturer. For example, all Boeing helicopters have common characteristics and there's a distinctive design philosophy in all Sikorsky machines. Aerospatiale helicopters have their own special French logic and the German engineers at MBB have their own way of doing things.

But regardless of their similarities, Bell 206s are different from Bell 214STs, Sikorsky S-58Ts are different from Sikorsky S-61Ns, MBB BO 105s are different from MBB BK 117s and so on (FIG. 11-1). To really get to know any particular helicopter, to get the real nitty-gritty, you must read the flight manual or pilot's operating handbook for that helicopter.

MBB Helicopters

Fig. 11-1. *Although similar in many ways, the MBB BO 105 on the left and the MBB BK 117 on the right have many differences. The flight manual for each must be studied carefully.*

For this reason, the following can only be a general introduction to the main systems common to most helicopters.

ENGINES

The heart of every helicopter is the engine or engines. There are two types of engines used in helicopters today: reciprocating (or internal-combustion piston) and turbine. Because you'll most likely be starting out in a helicopter with a reciprocating engine (FIG. 11-2) and because turbine engines are a subject onto themselves, I'll only cover reciprocating engines in this book. (To be honest, even though turbine engines can be highly complex and sophisticated, operating a reciprocating engine requires more skill than operating a turbine one. If you learn how to do the former, you'll have little difficulty moving up to the latter.)

The reciprocating engines used in helicopters are similar to the piston engines found in automobiles and trucks. The engines have cylinders and spark plugs, pistons and oil pumps, crankshafts and carburetors. Unlike automotive engines, aircraft engines are lighter and built with more expensive materials. Most are air-cooled, as opposed to liquid-cooled. Additional weight, complexity, and the risk of losing the liquid coolant are reasons why air-cooling is preferred in aircraft engines.

Another difference between automobile and aircraft engines is the ignition system. Instead of being comprised of a distributor, points, and voltage regulator, an aircraft reciprocating engine has magnetos.

212

Fig. 11-2. *Most helicopter pilots start their training in small helicopters with reciprocating engines, like this Schweizer 300.*

Magnetos

Magnetos always seem a bit mysterious to the new pilot. Fortunately, you don't have to know a whole lot about magnetos to use them and they're really not that complicated.

Magnetos are small, self-contained electrical generators whose sole function is to generate enough power to make the spark plugs spark. They are driven by a shaft taken off the engine so that whenever the engine is turning, so are the magnetos. Each cylinder in an aircraft engine has two spark plugs to provide for better combustion and more power. Each engine also has two magnetos that are independent of each other. Each magneto powers one spark plug in each cylinder so that if one magneto goes out, all the cylinders still get a spark.

The magneto switch simply opens and closes a circuit to each magneto; in the both position, the circuits to both magnetos are closed; right, only the right magneto's circuit is closed; left, only the left magneto's circuit. When the switch is in the off position, both circuits are open.

When the engine is not turning, switching on one or both magnetos does nothing, because the magnetos aren't turning either and are therefore not producing any electrical power. As soon as the engine turns over, the magnetos produce enough power to make a spark. If you have ever started an airplane by turning the propeller by hand, you know it doesn't take much of a rotation to start the engine.

If you switch to off while the engine is running, the engine will stop because, even though the magnetos would still be producing electrical power, the circuits to the spark plugs are now broken.

You check the magnetos on run-up by switching off each one separately and checking for a drop in engine rpm. A slight drop is normal because one spark plug in each cylinder is now not sparking. If the drop is excessive or if the engine stalls when you switch to one magneto, it's a sure sign that the magneto is bad because it can't keep the engine running by itself. The good one you just switched off was keeping the engine running.

You always want the redundancy of two magnetos. Never take off with one inoperative unless you absolutely must—for example, if you're on a volcano that was about to erupt. Otherwise, get the bad magneto fixed first.

Mixture Control

Automobile engines have a mixture control. It's one of the small screws on the side of the carburetor. If you've ever driven a car over the Rocky Mountains, you probably noticed the engine ran rougher and had less and less power the higher you climbed. This was because the lower air density at higher elevations resulted in your car engine's fuel-air mixture being too rich. People who live in high places have the carburetor in their cars adjusted to give a leaner fuel-air mixture than that of cars driven at lower elevations. A proper fuel-air mixture is the most important single factor affecting engine power output.

A too lean fuel-air mixture (too little fuel for the amount of air, by weight, entering the carburetor) will cause rough engine operation, overheating, backfiring, detonation, and loss of power. A too-rich mixture (too much fuel for the amount of air) causes rough engine operation and loss of power.

Because cars aren't often used over a wide range of altitude, a mixture control that can be adjusted with a screwdriver is sufficient. Aircraft need something more flexible; therefore, airplanes and helicopters with reciprocating engines are provided with mixture controls in the cockpit.

The mixture control normally has a red knob—an indication to the pilot to use caution when adjusting it (FIG. 11-3). The full rich position is with the control pushed all the way forward. Pulling the control toward you will cause the mixture to become leaner, meaning less fuel to air. Very delicate adjustments of the mixture control can be made by turning the knob, screwing it in or out, as opposed to pushing and pulling.

Generally, aircraft carburetors are adjusted for sea level operations. The correct position of the mixture control for start-up, takeoff, and landing is therefore the full rich position. At altitudes below 3,000 feet, there isn't usually much advantage to be gained by adjusting the mixture control, particularly if you're going to be making a lot of excursions down to lower altitudes.

If you plan to fly a long distance above 3,000 feet, however, adjusting the mixture control to the proper setting will provide a noticeable increase in power and decrease in fuel consumption.

Fig. 11-3. *The mixture control in this Enstrom F-28F is labeled PULL FOR IDLE CUTOFF—TURN TO LEAN. Note the key-operated magneto switch just above and to the left of the mixture control.*

Caution! Because a helicopter engine does not have a flywheel like an airplane engine, a helicopter engine will quit if the mixture is too lean. Some manufacturers recommend that the mixture not be leaned in flight. Be sure to consult the pilot's operating handbook of your particular helicopter to find out the proper use of the mixture control.

Carburetor Heat

Another thing aircraft reciprocating engines often have that automobile engines don't have is carburetor heat. Because the temperature of air passing through the carburetor can drop as much as 60 °F due to vaporization of the fuel mixture, moisture in the air will freeze and accumulate as frost or ice on the inside walls of the carburetor if the resulting temperature is below 32 °F. The restriction to the air flow causes a decrease in power and will eventually stall the engine.

The carburetor heater directs warm air into the carburetor to inhibit carburetor icing or to melt ice that's already there. Warm air is obtained by ducting outside air around piping in an exhaust muffler. The outside air is not mixed with exhaust gases, only heated by them. The carburetor heater control on the instrument panel simply opens and closes a butterfly valve that directs either unheated or heated outside air to the carburetor intake.

Because warm air is less dense than cold air, application of carburetor heat causes a slight decrease in engine power, which is evidenced by a decrease in engine rpm. If icing is already present when carburetor heat is applied, engine power should increase after 10 to 20 seconds as the ice melts away.

If your helicopter is equipped with a fuel injection system, it will not have a carburetor and, therefore, there is no need for carburetor heat.

Engine Oil System

An important part of any engine is its oil system. Oil is used for lubrication of vital moving parts inside the engine as well as for cooling. During operation of the engine, a pump located in the oil sump and driven by a drive shaft geared to the engine, pumps oil under pressure to bearings in the engine. The oil drains back to the sump, usually after passing through a radiator-type cooler.

An oil temperature gauge and an oil pressure gauge are provided to monitor the engine oil system. Of the two, the pressure gauge is the most important, but both should be checked often during the flight.

Low or no oil pressure indicates possible oil loss and subsequent engine seizure; an immediate landing should be made to investigate. High oil pressure could mean clogging of the oil filter or of an oil line. Engine failure is not as imminent as with low pressure, but the cause should be checked out soon.

Low oil temperature is usually due to low outside air temperature. If it's hot outside, it's probably an indicator failure. High oil temperature might be an indication of loss of oil, particularly if accompanied by low oil pressure. If the outside air temperature is high and oil pressure is normal, oil temperature will be higher than normal.

Engine Tachometer

The engine tachometer is similar to tachometers found in some automobiles and trucks. Like tachometers found in airplanes with constant-speed/variable-pitch propellers, tachometers in helicopters measure the rotational speed of the engine. And like the propeller control in such airplanes, the collective pitch lever in helicopters is the primary control of engine rpm during normal operation. (Some people argue that the throttle is the primary control of engine rpm; I'll explain this in a few paragraphs.)

Because the collective varies the pitch of the main rotor blades, which changes the power required to turn the blades, a special cam linkage, called a correlation device, is provided to correlate collective position with engine power. As the collective is raised, engine power is increased (in other words, the throttle valve progressively opens to allow more air into the carburetor). As the collective is lowered, power is decreased (the throttle valve closes). This correlation is not perfect, so the throttle control must be used to make up the difference in order to keep rotor rpm constant; therefore, the throttle control is considered the secondary control of engine rpm.

To make it easier to monitor both engine rpm and rotor rpm, most helicopters have dual tachometers that indicate both parameters on the same gauge (FIG. 11-4). Although the rotor turns much slower than the engine, the tachometer is geared so that both needles show the same indication (often indicated in percent) under normal conditions; therefore, as long as the engine and rotor rpm needles are matched in the green arc on the dual tach, everything is okay. Helicopters with two engines have triple tachometers: one needle for each engine and one for rotor rpm.

Fig. 11-4. *Single-engine helicopters have dual tachometers; this single-engine dual tachometer is in the lower left corner of the panel. Twin-engine helicopters have triple-tachometers.*

If the engine fails, the engine rpm needle will split off from the rotor rpm needle, and eventually wind down to zero. When practicing autorotations, autorotative flight is achieved only when the engine rpm falls below rotor rpm. If it doesn't split off, the engine is still giving some power to the transmission, even if the rate of descent is as great as in an actual autorotation. During practice autorotations, with engine rpm below rotor rpm, the throttle has full control of both engine rpm and manifold pressure and the collective has control of rotor rpm.

Too much throttle can cause engine and rotor rpm to go above the normal limits, particularly in low collective pitch settings. For this reason, most helicopters are equipped with engine governors that kick in and reduce engine output above a set rotor speed.

At high collective pitch settings, it's possible to reach the limit of engine power. If collective pitch is increased above this limit, rotor and engine rpm will begin to decrease or "decay." Try as it might, the engine simply cannot keep the rotor turning any faster.

Engine rpm cannot go above rotor rpm unless there is a failure of the freewheeling unit or the shaft between the engine and the transmission. If this happens, the transmission will be unpowered, even though the engine is still running, and autorotation must be entered immediately.

Manifold Pressure Gauge

The manifold pressure gauge measures power output. The gauge is an aneroid-type barometer that measures the air pressure in inches of mercury at a certain point in the induction manifold of the engine (FIG. 11-5). For any given engine speed, power output is proportional to the pressure inside the cylinders. The primary control of the manifold pressure is the throttle and the secondary control is the collective, although some people say it's the other way around.

When the engine is at rest, the manifold pressure gauge indicates ambient atmospheric pressure (FIG. 11-6). Theoretically, this is the maximum pressure the engine can produce on a given day, although in reality it will be two or three inches less. If the engine is supercharged or turbocharged, it's possible to obtain a manifold pressure above atmospheric pressure. A turbocharger pressurizes the air in a small, fast spinning turbine before directing it into the engine, which makes it possible for the engine to produce more power.

At idle speed the manifold pressure gauge will read quite low, usually 14 to 16 inches. The difference between the idle pressure and the atmospheric pressure is the maximum differential for that engine on that particular day.

You can check the accuracy of the manifold pressure gauge by comparing it to the pressure indicated in the Kollsman window of the barometric altimeter when you set the altimeter to zero altitude. If the altimeter is accurate and the manifold pressure gauge agrees with the Kollsman window indication, the manifold pressure gauge is correct.

Recall previous statements that the collective is the primary control of engine and rotor rpm and the throttle is the primary control of manifold pressure; also, the throttle is the secondary control of engine and rotor rpm and the collective is the secondary control of manifold pressure. Some sources on helicopter flying and many pilots say the exact opposite. Think of it this way.

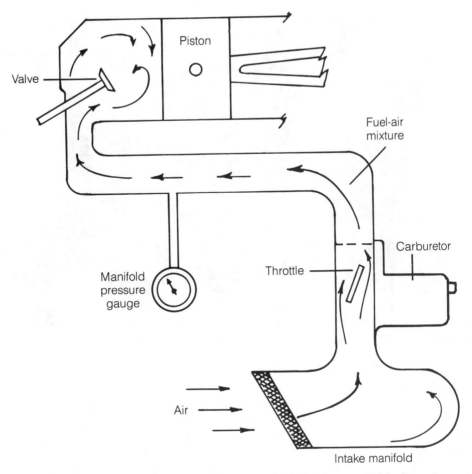

Fig. 11-5. *Location of manifold pressure gauge in the induction manifold of a reciprocating engine.*

In an airplane with a constant-speed/variable-pitch propeller, the function of the propeller control is to vary the angle of pitch of the propeller, similar to the way the collective varies the pitch of the main rotor blades in a helicopter. The instrument used to determine pitch is the tachometer. To set the pitch you want, you move the prop control until you obtain a certain engine rpm setting. As you increase the pitch of the propeller, the engine rpm decreases because the propeller blades are taking a bigger bite out of the air, producing more thrust and more drag. As you decrease the pitch of the propeller, engine rpm increases for the opposite reason.

The manifold pressure gauge is the main source of engine power information in an airplane with a constant-speed/variable-pitch propeller. The throttle is the control that varies the amount of air into the induction manifold and, therefore, manifold pressure. Increase, or open, the throttle and manifold pressure increases. Decrease the

Fig. 11-6. *When a reciprocating engine is shutdown, its manifold pressure gauge will indicate the outside ambient atmospheric pressure. In this Schweizer 300, the manifold pressure gauge (lower right corner) is indicating just under 30 inches.*

throttle and manifold pressure decreases. The same thing happens with a helicopter engine—manifold pressure is a function of throttle position.

Most helicopters have a correlation device that automatically opens the throttle as the collective is raised and closes it as the collective is lowered, but let's assume we have one without a correlation device for the moment. In such a helicopter, there is no connection between the engine and the collective. All you do when you move the collective is change the pitch angle on all the main rotor blades, just as the prop control changes the pitch on an airplane's propeller. As you raise the collective, rotor rpm decreases because of the increase in drag. Engine rpm also decreases because it is connected to the main rotor via the transmission. Lower the collective and both rotor and engine rpm increase.

Increase the throttle (power on) and manifold pressure goes up because it has to. You've opened the throttle valve more and now more air is drawn into the induction manifold. As manifold pressure goes up, so does the horsepower output of the engine. This, in turn, will lead to a rise in engine and rotor rpm, if collective pitch is held constant. Decrease the throttle and the opposite happens.

Now the difference. To get the best performance out of an airplane you use a high engine rpm setting for takeoff and a lower rpm setting for cruise. Helicopters, on the other hand, fly best at a constant rotor rpm setting. Large variations in rotor rpm—and engine rpm—are unacceptable.

In order to keep rpm constant in a helicopter, changes in collective pitch must be accompanied by changes in throttle. And because engine rpm and rotor rpm remain constant, the tachometer cannot be used to measure changes in the power setting. It is the collective pitch that controls rpm but its operation is hidden on the tachometer because of the requirement to adjust throttle whenever the collective is moved. The correlation device blurs the distinction between collective and throttle even more. Almost by default, the manifold pressure gauge has become the main instrument that indicates changes in engine power caused by both the collective and the throttle because both controls must always be coordinated.

Ask enough helicopter pilots and you will find some who say the collective controls manifold pressure and the throttle controls rpm and others who will say just the opposite. But all will agree that you really can't separate the functions of these two controls; each must always be used in conjunction with the other.

The interrelationships among collective, throttle, engine rpm, and rotor rpm, and manifold pressure cannot be overemphasized. The dual tachometer and the manifold pressure gauge must be analyzed to determine which controls to use and how much.

This does not mean you will always have to adjust both controls to obtain a change in rpm or manifold pressure. Because of their interrelationship each control can influence engine rpm and manifold pressure, as FIG. 11-7 illustrates.

PROBLEM		CORRECTIVE
Engine rpm	Manifold Pressure	ACTION
LOW	LOW	Increase throttle

Increasing the throttle opens the throttle valve, which increases manifold pressure and power output of the engine. The increase in power output, at constant collective pitch, causes engine and rotor rpm to increase.

Engine rpm	Manifold Pressure	ACTION
LOW	HIGH	Lower collective

Lowering the collective decreases the collective pitch angle of the rotor blades, which, at a constant engine power output, causes engine and rotor rpm to increase; however, the correlation device decreases engine power somewhat (by closing the throttle valve) so that manifold pressure also decreases.

Engine rpm	Manifold Pressure	ACTION
HIGH	HIGH	Decrease throttle

Decreasing the throttle closes the throttle valve, which decreases engine power output. Manifold pressure therefore decreases. Because collective pitch is held constant, engine and rotor rpm also decrease.

Engine rpm	Manifold Pressure	ACTION
HIGH	LOW	Increase collective

Increasing the collective causes blade pitch angle to increase, which means more power is needed to overcome the increase in drag. Engine and rotor rpm therefore decrease. The correlation device opens the throttle valve as collective is raised so that manifold pressure also increases.

Fig. 11-7. *Interrelationship among collective, throttle, engine rpm, rotor rpm, and manifold pressure.*

In most cases a coordinated application of both collective and throttle will usually be preferable. How much of which controls to use will be confusing at first, but after some hours it will become second nature.

MAIN TRANSMISSION

Connected to the engine is the main transmission, or main gearbox. (These terms mean exactly the same thing and are interchangeable. MGB is often used as the abbreviation.) The transmission is the heaviest single component of the helicopter, and one of the most critical. If the transmission stops, so does everything else.

The function of the main transmission is to transmit the power produced by the engines to the main rotor and tail rotor blades. This involves not only changing the direction of the rotational force, but also reducing the speed of the rotation.

The largest reduction in rpm is between the engine and the main rotor. Reciprocating engines rotate at about 3,000 rpm while the main rotor rotates about 250 to 400 rpm (these are only rough numbers). The rpm reduction is accomplished by a planetary and sun reduction gear system. The reduction from engine rpm to rotor rpm is expressed as a ratio. For example, if the engine rotates at 2,800 rpm and the rotor rotates at 350 rpm, the reduction ratio would be 8:1.

Tail rotors spin in the range of 2,500 rpm, so the gear reduction from engine to tail rotor drive is much less. Turbine engines rotate even faster, 20,000 rpm and above, so the gear reduction required by the transmission is even greater with a turbine engine.

All transmissions need oil. To supply the oil throughout the gearbox, at least one pump is needed and is usually located in the oil sump at the bottom of the transmission. The pump is driven by a shaft or gear connected to the gearbox itself so that oil pressure is provided whenever the transmission is turning. Cooling of the transmission oil is done by an oil cooler similar to a radiator found in a car. A sight gauge is always on the oil sump housing to permit oil level checks. The transmission oil level should always be checked during the pilot's preflight inspection.

Because of their high rotational speed, the engine-to-transmission input shaft, gears, and bearings are subject to the highest temperatures and most stress. It is essential that the input section is continually provided with oil for lubrication and cooling. The other parts of the gearbox require oil, too, but they can operate for a longer period of time without oil. For this reason, the emergency oil systems of some helicopters only provide lubrication to the input section of the transmission.

So far, helicopter designers have been unable to find a practical way to make a backup system for the transmission; therefore, the pilot is provided with several ways to monitor the condition of the gearbox (FIG. 11-8). The most important are two gauges that are found in all helicopters: the transmission oil temperature gauge and the transmission oil pressure gauge. Often these gauges are supplemented with low pressure and high temperature warning lights to catch the pilot's attention. Many helicopters also have magnetic chip detectors in the transmission and warning lights that

Fig. 11-8. *Boeing 234 pilots are provided with transmission oil pressure and temperature gauges and a switch that allows them to select and monitor each of the 234s five gearboxes. Transmission warning lights (XMSN) are also on the annunciator panel.*

indicate if the oil filter is clogged. Transmission fire indicators that consist of temperature sensors on the outside of the transmission are found on some helicopters. As mentioned in chapter 10, the newest helicopters are equipped with health and usage monitoring systems (HUMS) that measure the vibrations produced by the gears inside the gearbox. Changes in the vibrations give an advance warning of mechanical problems and possible failure.

The main transmission also transfers power to the tail rotor through the tail rotor drive shaft. The gears are fixed so that main rotor rpm and tail rotor rpm are always proportional.

Other items driven by the main transmission include electrical generators, hydraulic pumps, the transmission oil cooler fan, the torque meter pump, and rotor tachometer. Because each helicopter type is different, components powered by the accessory section of the transmission vary.

The rotor brake is not driven by the transmission, but is attached to one of the shafts driven by the transmission. Many small helicopters don't have the luxury of rotor brakes. On larger machines, they are a necessity. A rotor brake is required to slow down the rotation of the rotor blades after engine shutdown, and when engaging the rotors in high winds. The brake is also used when the helicopter is parked.

Most rotor brakes are similar to automobile disc brakes. The brake disc is mounted on either the tail rotor drive shaft or an accessory drive shaft. Pressure from a hydraulic system is used to force together the brake pad pucks, which are located on either side of the disc. The rotation of the disc slows and eventually stops the transmission, main rotor shaft, rotor blades, and tail rotor, because they're all connected.

CLUTCH AND FREEWHEELING UNIT

Some helicopters have a clutch and a freewheeling unit; others have only a freewheeling unit because it functions as both a clutch and a freewheeling unit. The purpose of the clutch is to make it possible to start the engine without it burdened down by the heavy load of the rotor system. The purpose of the freewheeling unit is to permit autorotation by freeing the rotor system from the engine.

Most helicopters use a centrifugal-type clutch or a freewheeling unit. There are several different types, but the basic operation is the same. The unit contains an inner shaft that is driven by the engine and an outer sleeve that drives the main transmission.

Between the shaft and the sleeve are either spring-loaded clutch shoes (like drum brakes in a car) or roller bearings. When the engine is at low rpm, centrifugal force is too low to overcome the spring tension of the clutch shoes; therefore, the clutch shoes do not press hard enough onto the outer sleeve to cause it to rotate. As engine rpm increases, the shoes gradually press harder and harder onto the sleeve until both the shaft and sleeve are rotating at the same rate.

If the unit has roller bearings instead of clutch shoes, centrifugal force causes the bearings to move up small inclined planes on the inner shaft and thereby exert increasing pressure on the outer sleeve. As the pressure increases, the rotation of the inner shaft is transmitted to the outer sleeve. When the engine and rotor rpm needles on the dual tachometer are matched (one needle superimposed over the other), the clutch or freewheeling unit is said to be fully engaged.

Some small helicopters have an idler clutch that requires manual operation by the pilot to connect the transmission to the engine during rotor engagement. With an idler clutch, the engine is started first with the clutch in the disengaged position. After the engine is operating at a sufficient rpm, the clutch is engaged carefully by the pilot and thereafter the transmission and rotors start to turn. During the normal shutdown procedure, the clutch is moved to the disengage position.

When the engine fails, the inner portion of the centrifugal clutch or freewheeling unit slows down. The clutch shoes or roller bearings come out of contact with the outer sleeve, which is now rotating faster than the inner shaft. The main transmission, main rotor, and tail rotor are free to rotate without having the burden of turning the inoperative engine, too.

MAIN ROTOR SYSTEM

The primary function of the transmission is to drive the main rotor shaft, or mast, which rotates the rotor hub, which causes the rotor blades to turn. The rotor shaft, the hub, and the blades comprise the main rotor system (FIG. 11-9). Integral to the main rotor shaft are subassemblies that are needed to transmit control changes to the rotor blades and devices that reduce the vibrations.

Fig. 11-9. *Comparison of two fully-articulated rotor systems from a second generation helicopter (top, Sikorsky S-61N) and a third generation helicopter (bottom, Aerospatiale AS 332). The reduction in complexity of the newer AS 332 rotor system is readily apparent.*

Rotor systems are classified three ways, depending upon how the rotor blades are fastened to the rotor hub. Blades in fully articulated systems can move about in three axes: *flapping*, *drag*, and *feathering*. A horizontal hinge permits flapping of the rotor blades (FIG. 11-10). Flapping helps equalize lift over the two halves of the rotor disc, as explained in chapter 2.

Fig. 11-10. *The flapping hinges permit the rotor blades to flap up and down in order to equalize the lift between the advancing blade half and the retreating blade half of the rotor disc: Enstrom F-28F main rotor head. (See also FIG. 2-8.)*

A vertical hinge permits the blades to move back and forth independently of each other. The hinge is called a *lead-lag*, *drag*, or *vertical hinge* and the movement is called *lead-lagging*, *dragging*, or *hunting* (FIG. 11-11). The ability to hunt is necessary to relieve the stresses that build up in the blades due to the coriolis effect on the blades.

The action of a rotor blade around the spanwise axis is called feathering and changes the pitch of the blade. Feathering is not something the blades are free to do, like flapping and dragging, but something the blades are directed to do by action of the pitch change rods. When you move the cyclic and collective controls, you change the pitch angle of the rotor blades around the feathering axis.

To be *fully articulated*, a helicopter must have three or more blades. The rotor system is *semirigid* if it has only two blades. In a semirigid system, the rotor blades are rigidly interconnected at the hub that is free to tilt with respect to the main rotor shaft (FIG. 11-12). The rotor blades can move within their respective drag and feathering axes, but they flap together as a unit.

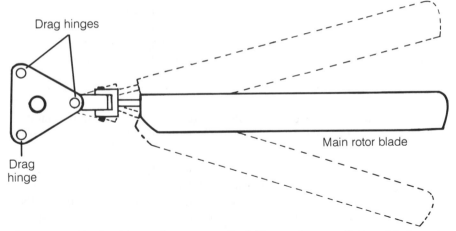

Fig. 11-11. *The lead-lag, drag, or vertical hinges allow each rotor blade to move back and forth independently of the others. The location of the hinge is chosen mainly with regard to controlling vibration.*

Fig. 11-12. *Two-bladed rotor systems are classified as semirigid. The stabilizer bars on this Bell 212 were invented by Arthur Young, designer of the Bell 47.*

With *rigid* rotor systems, the rotor blades can be feathered, but do not flap or drag about hinges. (Feathering is necessary or it would be impossible to control the helicopter.) Rigid rotor systems work because slight flapping and dragging is accomplished by the use of elastic materials in the blades and the rotor hub.

ROTOR BLADES

The main rotor blades are a helicopter's wings and are just as important to a helicopter as wings are to an airplane. Rotor blades create the lift that makes flight possible.

In contrast to airplane wings that are usually asymmetrical, helicopter rotor blades are symmetrical. This simply means that if you take a cross section of the wing, it is the same on the top as it is on the bottom, or, to be more technical, it is the same above the *chord line* as below it. The chord line is an imaginary line that joins the leading edge to the trailing edge of the blade.

Rotor blades are manufactured out of a wide variety of materials. The earliest helicopters used wooden rotor blades. In the 1950s and 1960s, metal blades became more commonplace. Today, the newest helicopters use blades made of composite materials (FIG. 11-13). To protect the front edge of the blade and extend the life of the entire blade, even composite blades have a layer of metal on the leading edge.

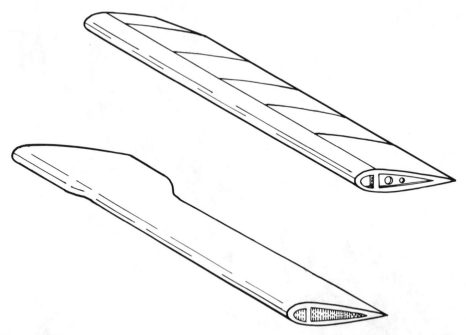

Fig. 11-13. *Comparison between design of metal main rotor blade (top) and composite main rotor blade (bottom).*

To reduce the weight without sacrificing strength, the core of metal blades consists of a honeycomb construction. Composite blades usually have foam on the inside. Some blades are pressurized with an inert gas and are equipped with indicating systems that show when the pressure of the gas has decreased. Lower pressure indicates a loss of gas due to a leak that might be caused by a crack in the blade. If a helicopter has blade inspection method (BIM) indicators, these should be checked on preflight.

SWASHPLATES

One of the most fascinating subassemblies of the helicopter is the swashplate assembly, which consists of a stationary swashplate and a rotating swashplate. Both swashplates encircle the rotor mast and are always parallel to each other. The purpose of the swashplate assembly is to transmit the linear, nonrotating control inputs of the pilot to the rotating components of the helicopter.

The stationary (lower) swashplate, also called the *stationary star*, is attached to the transmission and does not rotate with the main rotor shaft, but is free to tilt and move up and down the shaft. It is connected by control rods and scissor assemblies to the cyclic and collective controls. The rotating (upper) swashplate, or *rotating star*, rotates with the main rotor shaft and is also free to tilt and move up and down the shaft. Pitch-change rods connected to the rotating swashplate transmit movement of the swashplate to the main rotor blades.

When the collective is raised, both swashplates slide up the main rotor shaft, the upper one causing the pitch of all the blades to increase an equal amount. When the cyclic is raised, the swashplates tilt toward one side so that the blade pitch is changed on each blade cyclically. If that's as clear as mud, it's because it's hard to picture until you see it happen. Believe me, it works.

VIBRATION-REDUCING DEVICES

Vibrations are inherent in helicopters and selected engineers spend their entire careers trying to find ways to reduce rotary-wing vibrations. Compared to early machines, they've done quite well, but there is still much to be done.

The devices used on the rotor head include dampers, frequency adapters, and bifalars. Suspension bars, flexible mounting plates, elastometric bearings, and flexure assemblies are used to mount the transmission to the fuselage and reduce vibrations at the same time.

A pilot cannot do anything to these devices except check that they are properly fastened, if possible. On many helicopters, it's physically impossible to see the devices unless access panels are removed. The important point to remember is that an increase in vibration level or the onset of unusual vibrations could be caused by a problem with one or more vibration-reduction devices and should be investigated as soon as practicable.

FUEL SYSTEM

The fuel system of a helicopter is one of my favorite systems because it's one of the easiest to understand. Even in the most sophisticated helicopters, the fuel system is similar to that found in an ordinary automobile, except for some added gizmos and safety features.

The basic features of any fuel system are a fuel tank or fuel tanks, fuel lines to the engine or engines, fuel pumps, filters and strainers, and an indicating system.

Many different fuel tank configurations exist, even in helicopters of the same type. In a machine with only one fuel tank, all the features will be concentrated in that one tank. The tank will contain a drain valve in the lowest part of the sump, a device for measuring fuel quantity, a filter or screen, a filler cap, a vent, a shut-off valve, and often a fuel pump.

A system with several fuel tanks might have the elements distributed among the different tanks. For example, if one tank is situated above another and always feeds into the lower tank, the upper tank might have the filler cap and vent for both tanks while the lower tank has the drain valve and screen (FIG. 11-14).

Multiple tank fuel system have provisions for transferring fuel from tank to tank. Federal Aviation Regulations governing certification require that twin-engine aircraft have separate, independent fuel systems for each engine and a cross-feed mechanism that allows transfer of fuel from one system to the other in case of engine failure.

Fig. 11-14. *The two fuel tank filler necks (on either side of the cargo door) in the Aerospatiale AS 332 are used to fill up to seven interconnected fuel tanks below the floor of the aircraft.*

The drain valve is used to drain water and impurities from the fuel system, just like in an airplane. Water is heavier than aviation fuel so it should sink to the bottom of the fuel tank. Opening the drains is a part of every good preflight inspection.

Several strainers or screens are often located throughout a fuel system. Strainers filter out larger particles and are usually inaccessible to the pilot. Most systems also have fuel filters, some of which can be checked by the pilot.

Sophisticated fuel filters have bypass valves that allow unfiltered fuel to continue to flow to the engine in case the fuel filter becomes clogged. Unfiltered fuel is better than no fuel at all. These filters usually have some sort of mechanical indicating system, for example a small pop-out device, which becomes visible when the filter is clogged. On more complex helicopters, a fuel filter warning light illuminates in the cockpit when this happens.

Smaller helicopters usually have only one fuel pump per engine. The pump might be electrical or driven by a shaft from the engine. Larger helicopters often have one or more electrical boost pumps in the main fuel tanks in addition to the main engine-driven fuel pump.

The fuel quantity indicators are just like those in a car; however, measuring fuel quantity is not as precise as most people imagine. Particularly in helicopters, fuel quantity indications constantly fluctuate five to 10 percent due to aircraft movement, fluctuating more in very turbulent conditions. Changes in temperature affect fuel quantity and the gauges are never 100 percent accurate to begin with. So, take fuel quantity indicators with a grain of salt and always figure on a bit less than what the gauges show.

ELECTRICAL SYSTEM

At the Air Force Academy, we called our mandatory basic electrical engineering course "Black Magic 101." As far as I'm concerned, electricity still is black magic. As soon as any discussion of electricity goes past the analogy of electricity in a wire being similar to water in a pipe, I'm lost.

Fortunately, one doesn't have to be an electrical engineer to be an informed user of electrical systems in helicopters. I might never remember if either high voltage or high current kills you, but I do fine in a cockpit.

Helicopter electrical systems, even those of small helicopters, are more complex than most small airplanes, because the systems usually have direct current (DC) and alternating current (AC).

The major components of a DC power system include a battery, a starter-generator, a voltage regulator, relays, an inverter, and circuit breakers. The circuits are usually single wire with a common ground return. The negative terminals of the starter-generator and battery are grounded to the helicopter structure.

The main purpose of the battery in any aircraft is to motor the starter. Once the engine is running, the battery really isn't needed. Some small airplanes, like my 1946 Taylorcraft (had to get that in this book somewhere), don't have a starter but need a battery and electrical system to power the radios. To start the engine in my T-craft, all I have to do is spin the prop with the magnetos on.

All helicopters must have at least one battery, because the clutch or freewheeling gear between the engine and the transmission makes it impossible to prop start a helicopter. Even if you could engage the clutch and turn the rotors by hand, the weight on the engine from the transmission and the main and tail rotors would simply be too much for the engine. It would be like trying to push-start a car with the transmission engaged in first gear. You might be able do it, but you'd need a whole lot of people to get the car moving.

Helicopter batteries of choice are usually nickel-cadmium (ni-cad). Nickel-cadmium batteries, like sulfuric acid-type batteries in cars, are rechargeable; however, they hold the charge differently. Instead of being depleted in a more or less steady, straight line fashion like car batteries, nickel-cadmium batteries hold up to 80 percent of their amperage capacity for a longer period and then lose almost all of their charge in a very short period of time.

With a ni-cad battery, the engine won't go RRRRRRRRRR, RRRRRrrrrr, rrrrrr-rrrr, rrrrr, rrr, rr, r, r, —————, like it does when the battery dies trying to start a car on a cold morning. Instead, it will go RRRRRRRRRR, RRRRRRRRRR, RRRR-RRRR, RRRRRRRR, RRRRRRRR, —————. Once a ni-cad has lost so much charge that it can no longer motor the starter, it won't have enough charge left to power anything else either.

The battery is normally 24-volt and a certain rated amperage, for example 17 ampere-hour. The electrical system will normally consist of both direct current and alternating current circuits. The standard DC circuit is 28 volt and the standard AC circuits are 115 volt and 26 volt. Some helicopters are equipped with a 200 volt AC system.

The starter-generator functions as a motor and an electrical generator. When starting the engine, the starter-generator is powered by the battery and acts as a motor. After the engine is started, the starter-generator is rotated by the engine and produces 28 volt DC to supply the electrical system and keep the battery charged.

An inverter converts DC to AC. In some helicopters, vital engine instruments need AC power. Because it's obviously not a good idea to start the engine unless you can monitor what it's doing, these helicopters have a ground inverter that can be run off the battery. The ground inverter provides AC power to the engine instruments during the start. As soon as the transmission is engaged and the AC generators come on the line, the ground inverter can be turned off. The ground inverter is also an emergency backup AC system in the event the AC generators fail.

The 115-volt AC systems are needed to power autopilot and gyros. Also, 26-volt AC power is used for navigation equipment. Normally, 200-volt AC is only used to heat windshields or power rotor blade deicing systems. Some helicopters are equipped with transformer rectifiers that convert AC power to DC power.

All helicopters have provisions for external power. Some take only DC power, others have receptacles for both DC and AC. The standard plugs and receptacles for

DC and AC power are different to prevent inserting an AC plug into DC receptacle and vice versa.

If you use external power, be sure to follow the manufacturer's guidelines carefully because damage can be done to internal electrical components if switches are not in the proper position. It's also wise to find out precisely what the external power advisory light means for your particular helicopter. In some machines it means that external power is plugged in and turned on (which makes the most sense). In others it means that external power is plugged in, but could be on or off (if it's off, you might be draining power from the battery). In still others, it simply means the access door to the external power receptacles is open.

One last word about electrical systems: amperage (amps). The amount of power consumed by radios, lights, windshield heaters, and the like, is measured in amps. The rated power of a battery, when it is fully charged, is in amp-hours (AH). If a battery is rated at 16 amp-hours, this means it can power one 16-amp electrical consumer for one hour, or two 8-amp electrical consumers for one hour, or one 32-amp consumer for one-half hour. You get the idea. If the battery isn't fully charged, however, it will provide less than 16 amp-hours of power.

If the generators fail and the only source of power left is the battery, be very careful about which electrical components you leave on. As a rule of thumb, components that make heat use the most power, components that make light use a medium amount of power, and navigation and communication equipment uses the least amount of power. Reduce electrical consumption during an emergency by turning off unneeded components.

HYDRAULIC SYSTEM

Small helicopters do not need hydraulic systems; however, the bigger the helicopter is, the greater the dynamic loads on the rotor system become. Eventually, the loads become so great that it is humanly impossible to displace the flight controls without hydraulic boost.

For example, it's possible to fly a Bell 212 (max gross weight 11,500 pounds) with both hydraulic systems shut off, but it is considered an emergency maneuver and normally not allowed in training. In contrast, the Sikorsky S-61 and Aerospatiale AS-332L, both in the 19,000- to 21,000-pound range cannot be flown without at least one hydraulic system in operation. For safety reasons, both these large helicopters have two separate, independent systems. If one system fails, it's considered a serious emergency and a landing must be made as soon as possible.

Hydraulic systems are also used to operate many other components in large helicopters, such as landing gear, rotor brakes, and rescue hoists.

A standard hydraulic system consists of a reservoir, pumps, servos, tubing, and associated switches and pressure gauges. Because normal hydraulic pressures are in the 1,500-psi range, tubing and connections must be strong. All hydraulic lines should be inspected frequently because the most common cause of low hydraulic pressure is a leak in the system.

The hydraulic pumps are driven by shafts from the main transmission so that hydraulic pressure is always available even in the event of complete engine failure.

Hydraulic systems tend to be very complicated, particularly the servos connected to the flight controls and autopilot system. I spent hours studying a cross-section drawing of the auxiliary servo in the S-61 before I obtained even a glimmering of how it works. Twenty years later, the same drawing still looks like an impossible maze. I'm thankful that I don't need to know how it works to make it work.

The main thing you, as a pilot, need to know about hydraulic systems is how the controls should feel when the system is working properly, how they should feel when it is not, what the normal and abnormal instrument indications are, and what to do when something malfunctions. The best way to learn all of these things is in a simulator (FIG. 11-15). The second best way is from another pilot experienced in the machine, and to study. The second way takes much more time, but it's the only way available for most pilots. Unfortunately, there aren't too many helicopter simulators around.

Fig. 11-15. *The best way to learn about hydraulic failures is in a simulator: Bell 412/212 simulator.*

FLIGHT INSTRUMENTS

Helicopters have the same basic flight instruments as airplanes (FIG. 11-16), divided into three main categories: pitot-static, gyroscopic, and magnetic compass.

The pitot-static flight instruments are the airspeed indicator, the altimeter, and the vertical speed indicator. The gyroscopic instruments are the turn-and-slip indicator, the heading indicator (directional gyro), and the attitude indicator (artificial horizon). The magnetic compass is also called the whiskey compass.

You should be familiar with the operation of all these instruments if you have any previous flying experience, otherwise refer to one of the many fine texts on basic aeronautical knowledge.

Fig. 11-16. *Basic flight instruments in a Bell 212 (left to right): (top row) clock, barometric altimeter, attitude indicator, airspeed indicator; (middle row) vertical speed indicator, attitude indicator, triple tachometer, VHF homer (not a standard instrument); (bottom row) radar altimeter (behind cyclic), horizontal situation indicator, torque meter (primary power instrument in a turbine-powered helicopter).*

OTHER SYSTEMS

The larger and more complex a helicopter becomes, the more systems it has. A particular helicopter type might have some or all of the following systems: retractable landing gear, fire protection, heating and ventilation, ice and rain protection, lighting,

Fig. 11-17. *Studying aircraft systems does not stop after receiving a pilot certificate, continuing perhaps with a computer-based self-study curriculum or in a classroom: Bell Helicopter Customer Training Academy.*

automatic pilot, navigation, emergency flotation gear, life rafts, hoists, cargo sling. Military helicopters commonly have many other optional items.

Aircraft systems is one subject of two that you can never learn too much about. The other subject is flight regulations and procedures. Study and discussion about these subjects take up most of the professional pilot's training and study time, but its time well spent (FIG. 11-17). The more you know about your aircraft, the better equipped you'll be when problems occur.

12
Human Factors and Safety

If an aircraft accident occurs anytime, anywhere in the world, there is an 80 percent chance that, in the final analysis, it will be due to human factors.

Dr. Robert B. Lee
Australian Bureau of Air Safety

THIS IS THE MOST IMPORTANT CHAPTER IN THE BOOK. WHEN IT COMES right down to it, success as a pilot won't be due to your manual dexterity. What's in your head, your habits, your attitude, and your basic psychological makeup will make you or break you. That, basically, is what the subject of human factors is all about.

Dr. Lee's statement came from a research report, "Pilot Performance and Flight Safety," where he continued the thought: "This pattern has become firmly established over many years; however, particularly in airline operations, new technologies incorporating multiple redundancy and fail-safe concepts are becoming so reliable that, in future years, the proportion of human factors accidents may reach 100 percent, simply because the total, irrecoverable failure of machine components of the man-aircraft system will have been eliminated."

Dr. Lee's prediction hit me hard. It reminded me of the spacecraft computer in Arthur C. Clarke's science fiction novel *2001: A Space Odyssey*. At one point, the HAL 9000 computer confidently tells the human crew that no HAL 9000 has ever made an error. Even when there had been apparent errors, they were always caused by humans.

I'm pessimistic enough to believe that someday I'll be involved in a serious aircraft incident or even an accident (and optimistic enough to believe that I'll handle it properly and survive), but the thought that the probability is no less than 80 percent it will be my fault was hard to accept.

As much as I hate to admit it, Dr. Lee is probably right. The question is, how can we pilots improve the odds to our favor?

The first step, as Dr. Lee suggests, is to recognize the fact that human factors can and do effect pilot performance (FIG. 12-1). After we accept that human factors are a problem, then the next step is to find out everything we can about the subject. The final step is to take corrective measures to alleviate the adverse effects of human factors on our performance as pilots.

Because the cockpit is the pilot's workplace, most human factors problems are found here and the so-called pilot errors occur here. Unfortunately, our workplace is far from perfect.

Fig. 12-1. *Pressure to complete the mission often has a detrimental effect on pilot performance: Bell 206 LongRanger.*

Books have been written on human factors in aviation. A single chapter cannot cover the subject thoroughly, but I want to at least give you an overview.

A BRIEF INTRODUCTION

Most of the time, people adapt well to many of the design deficiencies in their working environments, even though their overall working efficiency might be reduced. The applied technology of human factors is meant to improve the efficiency of the system while providing for the well-being of the individual. When this objective is achieved, an increase in safety and efficiency of the man-machine interface will be realized.

The SHELL Model

The SHELL Model (FIG. 12-2) is one conceptual model of human factors. In the center of the model, is the human operator, or *liveware*. When working with a machine, the operator must contend with *software*, *hardware*, the *environment*, and other *liveware*. A mismatch anywhere in the system causes stress, which decreases efficiency and safety.

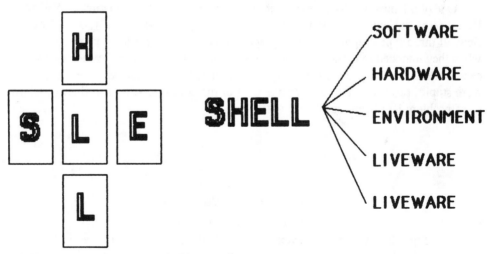

Fig. 12-2. *The SHELL model of human factors.*

Software relates to the machine itself and, with respect to aircraft, includes operating procedures, format of manuals, checklist design, language of information, symbology, and graphs/tabulation design.

Hardware, again in relation to aircraft, includes such things as controls, displays, warning systems, safety equipment, seat design, and cabin facilities.

Environment includes temperature, noise, vibration, humidity, pressure, light, pollution, and circadian/biorhythmic cycles.

Liveware includes leadership, communications, crew coordination, personal relations, and discipline.

Some fields of study used by human factor specialists are physiology, psychology, anthropometry, biomechanics, chronobiology, genetics, and statistics.

As you can see, the subject of human factors is very broad. Instead of trying to cover everything and being forced to only give you a brief taste of each area, I've decided to concentrate mainly on the liveware-hardware problems, the ones associated directly with the man-aircraft interface. For the sake of brevity, I've divided liveware-hardware human factors problems into two main groups.

The first group concerns the problems associated with the layout of the cockpit, and the switches and instruments themselves. In most cases, remedying the problems of "switchology," as I call it, is a relatively straight-forward task, although not necessarily an inexpensive one.

The second group concerns problems that originate more with the "man" than with the "machine." These are not so easily fixed, simply finding the cause of these problems might be extremely difficult; more on this later.

Switchology Problems

One of the most well-known switchology problems, discovered during World War II, concerns the control quadrants of the B-25, C-47, and C-82 (FIG. 12-3). Pilots who flew all three types reported that they accidently cut the throttle or mixture controls when they intended to reduce engine rpm with the propeller control. When safety officers looked into the problem, they quickly realized it was not because these pilots were stupid, but because of the arrangement of the controls. A typical liveware-hardware problem had been found.

Aircraft	Left	Center	Right
B-25	Throttle	Propeller	Mixture
C-47	Propeller	Throttle	Mixture
C-82	Mixture	Throttle	Propeller

Fig. 12-3. *Control quadrant sequence of three World War II aircraft.*

Let me give you a more up-to-date example of a switchology problem. This one in a helicopter.

To protect against an engine and rotor overspeed, the fuel control of the Aerospatiale AS332 Super Puma is designed to shut-down the engine automatically if the power turbine rpm go too high. Because the main conditions that can cause a power turbine overspeed (high speed shaft or freewheeling unit failure) happen so quickly, an overspeed warning light is provided so that the pilots realize the engine shut itself down due to an overspeed. This is good to know because one normally should not restart an engine that has shutdown for this reason.

A relevant point is that the overspeed light burns steadily when the engine is shut-down normally, but the light flashes when the overspeed mechanism shuts the engine down.

This creates a human factors problem. Most pilots, it seems, have a built-in aversion (although actually it is a conditioned response) to flashing lights in the cockpit. Their immediate gut reaction to a flashing lighted switch is to press the switch to make it stop blinking. A typical example is the master caution light in most aircraft. Some navigation systems, whenever there's a problem, flash a warning light that must be depressed before the pilot can begin corrective action. I'm sure there are numerous other good examples.

How can a flashing light that begs to be depressed be a problem? Consider the following scenario:

First, one engine fails due to an overspeed. The copilot sees the flashing over-speed light, says nothing, and then, unconsciously presses the overspeed switch to stop it flashing. (Many companies specify that the first action during any emergency procedure is to cancel the master warning light.)

A few minutes later, the captain, who has up to this point been concentrating on flying the aircraft, considers attempting an engine restart because he hadn't seen the overspeed warning and the light is now not flashing. The copilot, who canceled the only indication that would tell them they had an engine overspeed, can't remember canceling the light (because he did it without consciously thinking about it), and whole-heartedly agrees to a restart. The engine starts normally, because the broken engine-to-main gearbox shaft has no effect during the starting sequence; however, as soon as the pilots increase power, the unburdened power turbine shaft turns faster and faster. It's broken end rotates wildly and does untold damage to the input section of the gearbox.

The result? A carefully engineered warning system is rendered useless because a standard pilot response to a blinking light was not recognized by the cockpit designers. If such a thing ever happens for real, the cause of the accident will no doubt be put down as "pilot error." After all, the pilots did have an overspeed warning, didn't they?

The above scenario is not as far-fetched as it might seem. I observed it happen more times than I care to remember in the Super Puma simulator at Helikopter Service in Norway.

Switchology problems can be eliminated by redesigning the system. Although, it might take time to find the problem, in this case it took many hours in a simulator, and it takes time and money to modify the cockpit. Nevertheless, it can be done.

Even if we do eliminate all the switchology problems (and theoretically, this should be possible), it might be impossible to design a totally fool-proof (pilot-proof) cockpit because of the second group of man-machine interface problems that originate more with the man than with the machine.

PSYCHOLOGICAL BAGGAGE

Every pilot who steps into the cockpit of an aircraft carries a psychological flight bag of experience, background, and conditioned responses to outside stimuli. On the surface, we might all look like we're stamped from the same macho mold, but inside we are all very different. We are human. And despite the concentrated efforts of the training and operations departments to standardize our behavior in the cockpit, there will always be that element of unconscious psychological control that might cause us to act in a manner diametrically opposed to what even we ourselves know is correct.

Let me give you an example. When I first started flying helicopters, every once in a while, while hovering, I'd press the wrong pedal when I wanted to turn. Intuitively I knew I should press right pedal to turn right and left pedal to turn left, but sometimes a seat-of-the-pants reaction would cause me to press the incorrect pedal first, before I could catch myself doing it. "Why?" I asked myself.

Steering a machine with my feet was an unfamiliar action, especially after I'd driven a car for some years. But it was also vaguely familiar. "Were there other things I had steered with my feet?" I wondered. Then I remembered.

Have you ever gone sledding? If you sit on a sled, you have to steer it with your feet. To turn the sled to the left, you push your right foot forward; to turn it to the right, you push your left foot forward. Being from Pennsylvania, I did a lot of sledding when I was a kid, and in the stress of learning how to hover, every now and then, my unconscious mind would take over and tell my right leg to push the nose of the helicopter around to the left. It was a response I had learned years before I started flying. Psychologically, my emotional state when I sledded as a kid was probably not much different from my emotional state when I was learning how to fly. Both experiences were exciting, fun, and a little scary.

More examples: When you first learned to taxi a small plane, you probably had to concentrate to remember to turn with the rudder pedals and not the yoke, but after hundreds, or even thousands of hours, have you ever found yourself turning the yoke in the direction you want to turn while taxiing, even though your feet are doing the steering?

Or, have you ever caught yourself unconsciously pressing on the toe brakes of an airplane or helicopter when your final approach is a little too high and too fast? My 1946 Taylorcraft has heel brakes, an arrangement I never even knew about until I flew it for the first time. I have to continually remind myself to brake with my heels instead of my toes when I fly it, because my feet have more than 9,000 flight hours of toe brakes to unlearn. These are human factor reactions that most of us overcome with habit and experience during normal operations. When things start to get stressful, there's no telling what the unconscious might dredge up.

Compensating for every pilot's psychological baggage and stimulus-response habit patterns is impossible—all the more reason to get the human factor ambiguities, switchology problems, out of the cockpit. The only way is to standardize cockpits and procedures as much as possible and to employ an alert pilot force that is constantly on guard for man-machine interface problems.

I am convinced there are subtle human factor causes involved in most aircraft incidents and accidents. How can they not be involved? The very fact that whenever something out of the ordinary goes wrong the pilots are in a stressful situation is reason enough to suspect that their unconscious minds influence their responses to the stress-producing stimuli.

Sometimes, the problems of both switchology and psychological baggage are involved in the same incident:

A few years ago, a pitch-link on the tail rotor of an AS332L Super Puma broke while the aircraft was outbound on an offshore flight over the North Sea. The pilots made it safely to a ship after determining, correctly, there was something seriously wrong with the tail rotor. No one was hurt, there was minimal damage to the aircraft, the press reports were more accurate than usual, and everyone felt the pilots had done a good job. The investigation report concentrated on the pitch-link failure and maintenance procedures, and said nothing about human factors.

But listen to this. Shortly after the vibrations started, the copilot had suggested they should cut out the two autopilot lanes one at a time to see if the problem lay there. The captain, who was at the controls, looked down at the autopilot panel, and to his surprise, saw that both lanes of the autopilot were already disengaged. The copilot reengaged the autopilot lanes and they both worked normally. The point is, although the captain did not consciously remember cutting out the autopilot, he obviously must have.

There are two possible reasons why he might have done this: the first reason pure switchology; the second, more psychological.

In the first case, he might have been trying to uncouple the altitude and heading holds while still maintaining the autopilot, but accidently hit the wrong button on the cyclic, an error that probably every Super Puma pilot has made a few times (FIG. 12-4). This is a typical liveware-hardware switchology problem. The coupler release and the autopilot disengage buttons on the same control invites this kind of mix-up.

The other cause might have been rooted even deeper in the captain's unconscious. This particular pilot had more than 14,000 flight hours in helicopters at the time of the incident, 8,000 of these in the Sikorsky S-61. The automatic flight control system (AFCS) release button in the S-61 is located in the same position on the cyclic as the autopilot release button is on the AS332 (a sensible bit of standardization) (FIG. 12-5).

Because a yaw problem in an S-61 could be a hydraulic servo hardover, the first emergency action is to switch off the AFCS and the auxiliary servo. This procedure was a part of the captain's active conscious for more than 8,000 flying hours. It's not hard to speculate that his unconscious mind sent a signal directly to the ring finger on his right hand while his conscious mind was telling him there was a yaw problem with his aircraft. This might have been the reason he switched off the autopilot without even thinking about it.

You might wonder how a cockpit engineer could design for this kind of problem. To be fair, he probably can't. As I mentioned in the beginning, every pilot enters the cockpit with his or her own vast collection of experiences and subconscious reactions. This puts the whole shootin' match between the individual pilot and himseif.

Fig. 12-4. *The autopilot release switch in the AS 332 is the lower left button on the cyclic. The coupler release button (for the altitude and heading holds) is the top far right button on the cyclic.*

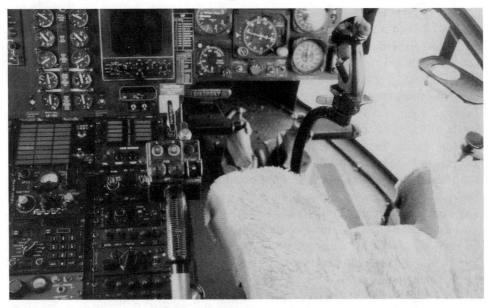

Fig. 12-5. *The AFCS release switch in the S-61 is the lower left button on the cyclic.*

So, what can we pilots do about it?

First, we must recognize that we are human machines and might not act under stress the way we want to or should.

Second, we can determine our subconscious responses to stress. A good session in a simulator will expose our gut reaction to numerous situations. Even on normal flights, you can be alert to your subconscious signals. For example, you shoot an unfamiliar approach to minimums—did you forget anything you should have done, timing at the outer marker or rechecking gear down with green indications? Whatever. Ask yourself why, then make a conscious attempt to change your habit patterns.

Finally, if you discover something in the cockpit that could cause a human factors-related accident, let the manufacturer know. Sure, we can all tough it out and make mental notes to ourselves not to get switches A and B mixed up, but remember this: If something strange has happened to you in the cockpit, if you've ever been confused about a warning light or indication or a switch position, then the same thing has probably happened to someone else, too. And it will no doubt happen again.

FUTURE PROBLEMS

The next generation of civilian helicopters, the ones you'll be flying during the 1990s and into the next century, will bring many welcome improvements from the pilots' point of view. Such things as four-axis autopilots, electronic flight information systems (EFIS) cockpits, increased endurance, and health and usage monitoring systems (HUMS) will become commonplace, if not required, and will all help make every pilot's job easier and safer. The goal of many helicopter operators is to duplicate airline safety records; many operators will achieve the goal in the 1990s.

New helicopters will also create new problems, not the least of which will be in the human factors area.

Autopilot-induced Deficiencies

Four-axis autopilots are available for helicopters and their use will certainly increase in the 1990s. A decent helicopter autopilot is a blessing welcomed by every hover lover, and I am continually amazed and impressed that a mechanical device can make a flying machine as complicated as a helicopter fly a coupled ILS approach and level off at 80 feet over the runway centerline, normally better than most pilots. But herein also lies its danger.

A good autopilot can fly the helicopter so well that pilots have a tendency to let "George" do the work most of the time, for the simple reasons that it's fun, it's easy, and it makes sense, safe sense, when the weather is bad; however, when pilots use the autopilot virtually all the time, on every flight, their flying skills suffer.

I'd be willing to bet that pilots who fly only four-axis equipped helicopters for more than a year are less proficient at basic flying skills than they were when they started, unless they make a concerted and continuous effort to fly without the autopilot coupler functions engaged from time to time.

I can support this viewpoint from experience as a former AS332 flight and simulator instructor who has flown with and trained pilots from Sealand Helicopters, ERA, Pelita Air Service, the Japanese Defense Force, the Korean Air Force, Royal Flight Oman, Lufttransport, Helikopter Service, and others.

One might argue that airplane drivers have had autopilots for many more years than helicopter pilots and that their skills have not suffered. I wonder. Even if it is granted that 747 captains are as good as or better pilots than the old stick and rudder aviators, I would argue that the average helicopter pilot needs to retain his basic flying skills more than an airplane pilot, for hovering alone, if for no other reason.

The solution is simple and obvious. Pilots should fly often without the higher order of autopilot coupler functions (altitude, airspeed, heading holds, and the like) engaged. Lufttransport, before it was merged with Helikopter Service, required its pilots to fly "coupler out" at least one flight every work period. This was a good policy and the helicopter operators of the 1990s would be wise to adopt a similar procedure for all pilots who regularly fly autopilot-equipped helicopters.

HUMS Unknowns

In the simplest sense, health monitoring systems for engines, in the form of oil pressure and temperature gauges and magnetic chip detectors, have been around for decades. The British Royal Air Force's famed Red Arrows aerobatic team reportedly has a meantime between engine failures of 10,000 hours, thanks, in part, to a HUMS. The EH 101, a three-engine helicopter jointly developed by Westland and Agusta, has a HUMS as standard equipment (FIG. 12-6). Aerospatiale is also working on HUMS and is closely involved with Bristow Helicopters and the British CAA in the testing of a HUMS installed in a Bristow AS332 Tiger. Not surprisingly, the AS332L Mk II incorporates a sophisticated HUMS.

New developments in vibration detection and analysis and, of course, computer technology are making it possible to detect such things as a failing roller bearing in a turbine engine or a hairline fracture on one tooth of a planetary gear in a helicopter transmission simply by measuring the change in vibration. With experience, the systems might even be able to determine that the failure will come in x number of hours, plus or minus a few.

This might turn out to be a mixed blessing for operators and pilots. Imagine, for example, a situation in which the HUMS says the tail rotor gearbox will seize in 10 hours, plus or minus 5, with a statistical probability of 99 percent and the aircraft is scheduled for a 4.5-hour flight. Should the flight go or not? What if the flight is to be only four hours? Three hours? Thirty minutes? Where does one draw the line?

The problem of the 1990s will be that everyone, pilots, maintenance engineers, managers, and manufacturers, will be inexperienced with HUMS.

Pilots will be understandably conservative and wary. If there's a HUMS warning they'll want to know what it is and why.

E.H. Industries Ltd.

Fig. 12-6. *The EH101 will have the most advanced health and usage monitoring system (HUMS) of any helicopter in the world when it becomes operational.*

Maintenance and operation managers will be understandably suspicious. Does that gearbox really have to be changed now just because the HUMS picked up an unusual vibration? What if the HUMS is unserviceable? Will it be, should it be, a minimum equipment list item?

Manufacturers will be hard pressed to prove that the HUMS really does what it's designed to do. Cost savings are being touted as one of the advantages of HUMS, but initially, the devices might end up costing operators and manufacturers more than if the HUMS had never been installed at all.

Basically, it's going to take time before the industry works these things out and everyone becomes comfortable with HUMS. It will happen because HUMS has the promise of increased safety for the non-redundant items on helicopters. The 1990s will be the proving period.

Finger Trouble

Any pilot who has worked with an advanced autopilot knows that the most frequent mistakes made by pilots, even after they know how the system operates, are:

- Pushing the wrong buttons at the right time.
- Pushing the right buttons at the wrong time.
- Pushing the right buttons in the wrong sequence.
- Thinking that an autopilot function is off when it is on.
- Thinking that an autopilot function is on when it is off.

In other words, finger trouble.

One of the reasons for this, if not the primary reason, is the manner the autopilot functions are displayed. Usually, the annunciator lights are shown on one central autopilot panel that is often on the center cockpit console (easy to reach, but out of sight). Sometimes the annunciators are duplicated elsewhere in the cockpit, on the panel in front of the pilots, or on flight instruments. For example, airspeed hold might be displayed on the airspeed indicator, altitude hold on the barometric or radar altimeter, localizer and glide slope hold on the horizontal situation indicator or artificial horizon.

Of the three methods, the last one is the best because this is the method the autopilot annunciators will most likely and most often be seen by the pilot. The simple reason is that the flight instruments are an integral part of every experienced pilot's cockpit scan, the instrument cross-check that never seems to be quick enough during a checkride. The autopilot annunciator panel on the center console is not a frequent part of most pilots' instrument cross-check. Hopefully, future cockpits will be designed around the principle of putting an indication of each autopilot hold function on the related flight instrument, a relatively simple task with EFIS tubes in the cockpit.

I have no experience with EFIS, but these glass cockpits provide virtually endless possibilities for displaying just about anything and everything that could be displayed in front of the pilots and this in itself will create confusion. The more choices available to do a certain task, the greater the chance for mistakes and misinterpretations.

Even something as simple as a distance measuring equipment (DME) receiver might be confusing when it is possible to monitor up to four separate DMEs on the same indicator (with dual VOR receivers that have DME hold functions). Accidents have occurred because the pilots misinterpreted their position based upon information from a DME receiver that was incorrectly tuned to the wrong transmitter.

Some area navigation systems have so many display possibilities that it's impossible to use the system without reading the manual every time; most pilots learn the bare basics of the system and just ignore the rest. The only practical solution to cut down the confusion element is to standardize company procedures so that switches and displays are normally set in a certain way. This might sound almost sacrilegious to the designer who has worked nights to cram as many microchips as possible into the black box, but in the cockpit, things have to be clear, simple, and straight-forward.

Finger trouble might become one of the most frequent human factor problems for the next decade's helicopter pilots.

Electronic Glitches

Along with autopilots, HUMS, EFIS, and other advanced systems will come electronic glitches. We have these problems already with the present generation of the AS332s, the Boeing 234s, and the Bell 214STs.

We seem to be on an electronic threshold, not just the helicopter and fixed-wing industries, but the entire modern world. Engineers can design intricate and complicated systems and manufacturers can and do build them, but many things are not yet constructed to a reliability that guarantees perfect operation at all times, and in all environmental conditions.

British Caledonian Helicopters had numerous spurious engine malfunctions with their Bell 214Ss until they realized the problem was rain water running down the wires to a connector plug on the engine casing. The AS332 often gives incorrect landing gear indications because of problems with the landing gear position switches. Disconnection of an engine fuel filter warning light on a Helikopter Service Boeing 234 (because it came on too often) eventually resulted in the filter clogging, the engine seizing, and control problems that caused the grounding of the machine for over six months.

Glitches and bugs take time to find and sort out. An electrical fire can wreak havoc with virtually every system in a machine if redundancy and wiring are not properly thought out. The space industry uses the fail-safe philosophy: If it can fail, it will, and therefore a safe backup system is designed in. And even with this fail-safe principle, the space shuttle still malfunctioned, disastrously, partly because of an environmental condition.

Helicopters are used and needed all over the world, from the hottest deserts of Saudi Arabia to the coldest Arctic glaciers. Electronics and computers offer technological progress, but must be able to withstand the worst weather imaginable.

Discomfort and Boredom

It used to be that helicopters were very short-range vehicles. This is no longer true because in many parts of the world, four- and five-hour helicopter flights are not uncommon, particularly in the offshore business. As the exploration for oil and gas moves farther out to sea, the flights to the rigs and platforms will become longer and longer (FIG. 12-7).

Autopilots make these long trips easier for the pilots, but there are two things autopilots can't help: physical discomfort and boredom.

Even the best pilot seats available today become uncomfortable after a pilot has been sitting for more than three hours in the cockpit. The vibrations inherent to every helicopter only exaggerate the problem. Obviously, the positions of helicopter controls and the pilot's need to sit in a certain position to manipulate the controls create unique constraints on seat construction.

There must be a better way because long-distance truck drivers have better accommodations than most helicopter pilots. Luxury cars boast ergonomically designed seats. Perhaps the cockpit designers from Aerospatiale, Boeing, and EH Industries need to make visits to Stuttgart, Turino, Munich, Toulouse, Güteborg, Toyota City, even Detroit (FIG. 12-8).

Boredom will also come as helicopter operations, particularly those over water, become more like airline flying. Boredom already is a problem on long routes, but the longer flights of the 1990s and improved autopilots will give the pilots less and less to do while en route.

Fig. 12-7. *As exploration for offshore oil moves farther and farther out to sea, helicopter flights will become longer and longer: Ekofisk Hotel platform, North Sea.*

Fig. 12-8. *This specially-designed pilot seat in a Helikopter Service S-61 is one of the most comfortable in the industry, but even it becomes uncomfortable after three or four hours of flight time.*

I don't know the solution. As with the problem of proficiency loss due to increased usage of autopilots, perhaps we should seek solutions from our fixed-wing brethren. I do know it's hard to remain professionally alert when everything is working and you're droning along at 5,000 feet in solid cloud for two to three hours.

The issue needs to be taken up and brought out into the open. Ideas should be solicited and solutions sought. What about incorporating computer games into the EFIS? Let the pilots play chess or Nintendo between reporting points. Or how about showing a video film on one of the screens? Or perhaps a series of training films about aircraft systems? Northwest Airlines has already introduced seatback-mounted video systems on a Boeing 747. Do the Boeing 747 pilots have one in the cockpit, too? How do they stay alert on eight-hour legs?

As radical as these suggestions might seem, something will have to be done. Many pilots read newspapers and magazines in the cockpit and this, I think, is preferable to staring straight ahead in a mild hypnotic trance or sleeping. It's not easy to remain alert when you can't keep your eyes open.

Greater endurance capabilities of the 1990s helicopters will make in-flight boredom and seat-of-the-pants discomfort a reality for helicopter pilots in the future.

ADVICE TO THE MANUFACTURERS

President Ronald Reagan once said, "The Space Invader-playing kids of today will be the fighter and bomber pilots of tomorrow."

Even though he didn't mention helicopter pilots, they certainly should be included in the group of future pilots. What's more, the former president's prediction is already coming true. How will this affect human factor problems in the cockpit?

Not more than 10 or 15 years ago, the space and aircraft industries were the epitome of high-tech. In many ways, they still are, but since the advent of inexpensive microchips, "smart" machines are now commonplace in most homes. The gap between sophisticated aircraft and sophisticated household machines has narrowed. An entire house can be controlled by a central computer. The Electronic Industries Association/Consumer Electronics Group in late 1988 announced a new wiring standard called the *consumer electronics bus* that will enable microprocessor-equipped appliances built by one company to communicate with those built by another.

This means that more and more people will use sophisticated electronic and computer-controlled devices on a daily basis. Many children learn to operate electronic equipment before they can read. At age four, my youngest son knew how to operate the remote controls to our video cassette recorder and television, find and play games on our Macintosh computer, use various cassette players, and heat food in a microwave oven. Operating electronic equipment is second nature to him.

Aircraft designers will have a new human factor element to consider. Instead of the automobile, electronic, and other industries mimicking the designs of equipment found in aircraft, the aircraft manufacturers might find themselves copying panel designs from these industries in order to avoid human factor problems in the cockpit.

This is not to say that aircraft will loose their place on the cutting edge of technology, but that aircraft designers will have to be more aware of the designs of equipment made by other industries.

In the past, pilots had to contend with transfer of learning problems between their airplanes and their automobiles. These problems will seem minor to the pilots of future generations who will have to contend with transfer of learning problems between their aircraft and their cars, their computers, their home entertainment systems, and numerous other gadgets, appliances, and machines, some of which have yet to be invented.

There will be international "standards" developed and accepted, sometimes by agreements and official decrees, but perhaps more often by the company that is able to sell the most of a particular product first. If a similar machine does not fit the accepted norm or the standard that people have become accustomed to, problems will arise.

This was illustrated when Helikopter Service installed a new security system that required the use of magnetic-strip identity cards to enter a building. The old system required insertion of a card and punching in a four-digit code before the door would open. The new system required that the code be punched in first, then the card inserted. If you inserted the card first, a red light blinked indicating something was wrong.

On the first day, only a few people could get in the building. Even though instructions had been sent to everyone, most didn't bother to read them. They assumed they should insert the card first and then punch in their code. When this didn't work, they figured there was something wrong with their card or that they had the wrong code.

The system was the problem. The fault was that of the designer who had not considered that a standard for card-and-code door opening systems had already been accepted.

There might have to be a radical change in the way aircraft are designed. In the past, the machine was foremost. The goal was to make the machine work and if a switch or lever was in an awkward position for the pilot, then he just had to adapt. This has changed a great deal since World War II and aircraft designers spend much more attention to ergonomic factors inside the cockpit.

First Point

Designers of aircraft must also look outside the cockpit, at the numerous other sophisticated machines that are becoming commonplace, or are already commonplace, when considering human factors implications in the cockpit.

Everything in the cockpit will have to be examined with human factors in mind. From the simplest mechanical considerations, such as the way the seats are adjusted, to the most sophisticated computer systems. Designers will have to stay up-to-date with currently accepted standards in the outside world.

Are pull-down menus on computers so widespread that they can be considered a standard that should be used in the cockpit? Should the QWERTY typewriter keyboard, like that used with virtually all personal computers, be the standard for aircraft navigation and computer systems (FIG. 12-9)? Should the artificial feel in a fly-by-wire control stick have the same feel as a Nintendo joystick? Should the clock be digital or analog, or both? These are questions that must be constantly and continually asked and addressed.

Fig. 12-9. *Should the standard QWERTY keyboard of typewriters and computers be used in aircraft of the future? Aircraft designers must consider human factor problems before answering this question.*

Great strides in cockpit design have already been made, of course, and manufacturers spend a lot of time and money in this area. At the 1985 Paris Air Show, Lockheed exhibited a simulator that the company's human factors engineers are using to design the cockpits of the future, particularly with respect to new avionics, EFIS, touch screens, voice actuation, and many other innovations currently being researched.

Optimum cockpit design for any aircraft will not be found by the manufacturer alone, no matter how many test pilots and line pilots are consulted during the design process. The manufacturers will, no doubt, get their final product close to perfection, but it takes operational experience with an aircraft before all the kinks are ironed out, or even found out. This is where the normal line pilot can and should make an input.

Second Point

Manufacturers must establish, promote, and use an effective feedback system so that ideas and suggestions from line pilots can be obtained.

Never in my career as a pilot have I seen such a system from any manufacturer. Perhaps a system exists, but if it does, the word has not gone out to the people that are flying the machines every day.

United Technologies Sikorsky Aircraft

Fig. 12-10. *Manufacturers should actively solicit ideas from the pilots and mechanics who actually operate and maintain their aircraft to find out ways to eliminate human factor problems: Sikorsky SH-60B SeaHawk.*

Every successful company believes it is "the man on the shop floor" who best knows how to do his job and who has the most useful suggestions about how to do it better (FIG. 12-10). Good companies solicit information from every level.

In my experience, aviation companies are often very conservative and many even have military-like organizational structures. Information in military hierarchies goes up and down the chain of command, but it usually flows down a lot easier than it goes up. If the chief pilot or chief of maintenance does not agree with a line pilot's or mechanic's suggestion, the idea stops there and never gets to the manufacturer where it might have been accepted. The only exception is in the case of an accident. Then the suggestion is heard.

A reporting system connecting line pilots directly to manufacturers would be an excellent way to receive feedback about present and future cockpits. If Dr. Lee's predictions about the proportion of human factor accidents reaching 100 percent is correct, then constant awareness of and attention to human factor problems will be the only way to prevent aircraft accidents in the future.

Conclusion

The 1990s will see many helicopter operations becoming more and more like fixed-wing airline operations. The period will bring new challenges and technologies that will require new thinking and novel solutions to overcome the human factor problems associated with them.

Perhaps it is impossible to eliminate all human factor kinks in aircraft, but every cockpit can be improved. Manufacturers spend a lot of time on cockpit design, but they don't have the experience of hundreds or thousands of hours of normal operations that line pilots get on a day-to-day basis.

If nothing else, the people in the helicopter industry are flexible and resilient. They have met such challenges head-on in the past, and I'm sure they'll tackle them in the future. We, as pilots, are morally responsible to give them the feedback they need.

In the final analysis, the human machine might never reach the multiple redundancy and fail-safe reliability of supersophisticated aircraft, but we sure can do a lot better with human factors problems than we are doing today. Besides, we really shouldn't depend too much on machines, no matter how sophisticated they become.

Remember what happened in *2001: A Space Odyssey*. Even a HAL 9000 computer can make a mistake.

13
Born-again Copilots

A superior pilot is one who stays out of trouble by using his superior judgment to avoid situations which might require the use of his superior skill.

Directorate of Flight Safety
Royal Air Force

I<small>F YOU PURSUE A CAREER AS A HELICOPTER PILOT, YOU WILL EVENTUALLY</small> find yourself in the position of "born-again" copilot. This isn't a religious thing, but part of normal career progression.

I wrote the following article for *Vertiflite*, the official publication of the American Helicopter Society. Hopefully anyone aspiring to become a professional helicopter pilot will find it useful.

ONE STEP BACKWARD, TWO STEPS FORWARD

Moving up a career ladder is standard in any profession. For the professional pilot it can mean moving from single-engine piston helicopters and airplanes to turbine-powered aircraft to twin-engine machines and eventually to even bigger multiengine craft. Certificates progress from student to private to commercial to airline transport pilot.

Job progression in the cockpit is generally from copilot to captain. This is often considered the natural order of an aviation career. Much ado is made along each step of the way: more stripes, more responsibility, more pay.

But moving up is not always a straight-line progression. Sometimes a pilot takes one step back in order to take two steps forward later. Often a pilot moves from a position with ultimate authority and responsibility to a subordinate position with less authority and responsibility. In other words, sometimes during his or her aviation career, the typical pilot becomes a "born-again" copilot.

This captain-to-copilot transition is not as unusual as it might seem at first glance. Think, for example, of an airline pilot upgrading from the left seat of a Boeing 727 to the right seat of a Boeing 747 or of a helicopter pilot moving up from right seat of Bell JetRanger to the left seat of a Bell 214ST (FIG. 13-1). Pilots who change companies often find themselves in the copilot's seat, at least for a few flight hours anyway. Most companies routinely schedule low-time captains with high-time captains as copilots to give the low-timers more pilot-in-command time.

Not much thought or discussion or training is given to whatever problems are associated with born-again copilots. It's generally assumed that good captains also make good copilots. I don't agree.

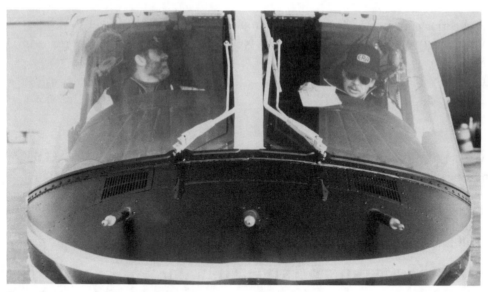

Fig. 13-1. *An experienced Bell 206 pilot might find himself as a copilot in a Bell 214ST.*

PASSIVE COPILOTS

The main problems with born-again copilots are not so much physical as they are psychological. The problems have more to do with a person's perception of the role of being a copilot rather than his or her ability to fly the aircraft. In my experience, there are two types of born-again copilots and different problems associated with both types.

The first type of born-again copilots includes those pilots who have recently changed companies or aircraft and find themselves flying as copilots on a regular basis for the first time in a long time.

The biggest problem with this copilot is passivity, caused, in part, by a feeling of inferiority. Few pilots would admit they ever have an inferiority complex, but if they are at all honest with themselves, I think most pilots will confess to some feelings of inadequacy when confronted with a new aircraft or operation. (There is, after all, only one Chuck Yeager.)

We might hide our true feelings behind a facade of well-practiced bravado—we do have our images to uphold, don't we?—but deep inside there's one, the knowledge that we don't know everything we should know and two, a very strong desire not to screw up.

So what happens? A company hires an experienced pilot with bags of pilot-in-command time or promotes one of their prize Sikorsky S-76 captains to the left seat of their Aerospatiale AS-332. He's good and he's already proven himself.

They check him out in the aircraft, perhaps send him off somewhere for expensive but worthwhile simulator training, schedule an instructor to fly with him for a while, and then put him on the line. For most practical purposes, this pilot is as ready as he'll ever be. He can fly the machine, read the charts, handle the radios, outwit the autopilot, run the checklist, and file the flight plan like the pro that he is.

If he really is a pro, he knows he's still a step or two behind most of the pilots who have more time in the aircraft or with the company. He has an inferiority complex with respect to those pilots who are more experienced in the job he's just begun to do. This is healthy, in a way, because in some respects he is inferior. But the point is, when the chips are down, this guy is going to be more prone to acquiesce to the decisions of a captain whom he perceives as having superior knowledge.

NOT THE RIGHT STUFF

Unfortunately, I speak from experience. I had more than 1,600 total hours and 1,000 hours in type when I left the Air Force for my first civilian flying job. I had flown as aircraft commander on some pretty hairy rescue missions in Iceland and Alaska. Offshore flying would be a piece of cake, I thought.

It wasn't. Wake-ups at 0200, instrument takeoffs from rigs at night in the middle of snowstorms, and seven to eight hours of continuous shuttle flying with 20 or 30 landings and breaks only for hot refueling were quite a change from the almost leisurely pace of military flying. During that first year, I flew more than I had averaged in two years in the Air Force.

I flew as copilot with captains who already had years of offshore experience, captains who really knew what they were doing. I was with one such captain when I realized I was not as good a copilot as I thought I was.

We had just taken off for an early morning flight to an oil field (FIG. 13-2). Shortly after I had finished reading the after takeoff checklist, the engine oil pressure low light blinked on. We simultaneously checked the gauges immediately: The number one engine oil pressure had dropped to just below the redline.

Fig. 13-2. *Shortly after takeoff, we noticed that the engine oil pressure low warning light had come on.*

The captain decided to return for landing and instructed me to retard the malfunctioning engine to ground idle. Even though it was good VFR and the runway was 7,000 feet, he decided to keep the engine with low oil-pressure running at ground idle, just in case it was needed for a go-around.

In the back of my mind, I remembered something about the loss of engine oil pressure and the inevitability of the engine seizing soon after all oil is lost. I thought I remembered the book procedure for an engine oil low pressure light was to shut down the engine immediately. Although it did not seem that unreasonable to me to leave the engine at ground idle, I couldn't remember anything in the book about such a procedure. "Well," I thought, "it's possible I've forgotten the correct procedure or got it mixed up or never knew it well enough in the first place. Besides, this captain has much more experience than I do, and he's probably right. Who am I to question him?"

On the other hand, I did recall once seeing an oil-starved engine seize and catch fire in an Air Force simulator. I really hoped that wasn't going to happen (if it did we still had the fire bottles). I really thought I should say something. But I just couldn't do it. I kept my mouth in the shut-off position, even though I was fairly certain that was what we should have done to the engine.

The engine kept running, temperature normal and steady, pressure steady but below the redline, during the captain's finely executed landing. We shut the thing down while taxiing to the hangar. Fortunately for the company, the mechanics, and ourselves, some oil remained in the engine and no damage was done by our failure not to shut it down sooner. We got another aircraft and took off.

On return, we had an appointment with the chief pilot. He explained, in no uncertain terms, that the correct procedure in the event of a loss of engine oil pressure is, as I had thought, to shut down the engine as soon as possible.

I realized that day that I had failed my "captain-to-copilot" transition and I had taken the wrong approach to being a born-again copilot. I wrote in a notebook: "Be more aggressive. When you know something, or are even fairly sure of something, express it. If the captain does something wrong, tell him. You must not allow yourself to be caught in an unsafe or incorrect situation due to the improper action of the captain."

From that day on I became more expressive and assertive about things happening in the cockpit. For a while, it was difficult to point out mistakes to some captains, even tactfully. Let's face it, there are some people who just don't like being corrected. But I did it, and it usually wasn't as hard as I thought it would be.

Now that I do most of my flying from the captain's seat, I try to impress upon my copilots, especially those with much less experience than myself, that they should not be afraid to point out anything that I am doing unsafely or incorrectly.

CAPTAIN/COPILOTS

Which brings us to the other type of born-again copilots. These are captains who occasionally find themselves flying as copilots in aircraft that they usually command. The biggest problem with these "captain/copilots" is not passivity, but apathy.

Because the "heavy mantle of responsibility" has been lifted from their shoulders, captains who fly as copilots might feel they can now take it easy and let the other captain do the worrying for both of them. As in the case of the passive copilot, the cockpit is being managed by only half the team.

New captains soon learn there are some senior captains they would rather not have as their copilot because the young captain often ends up doing both jobs, captain's and copilot's. Senior captains, I suspect, become so used to being assisted by other crewmembers that they forget how to be good assistants themselves.

If one adds apathy to the out-of-practice captain/copilot, the situation can really get interesting. Unlike the passive copilot, the captain/copilot would probably have few reservations about pointing out irregularities to his cockpitmate. The question is whether or not he would be alert enough to notice the irregularities in the first place.

With respect to safety, or rather lack of it, it's probably a toss-up between the passive copilot and the apathetic captain/copilot. I wouldn't put my money on either. The captain/copilot would possibly have more experience than the passive copilot, but this wouldn't count for much if he is too laid back to use it. Worse, the captain/copilot might cause the other captain to relax, too.

On one side of the cockpit, you have the captain thinking, "Since my copilot today is captain-qualified, I don't have to watch him as closely and I can take it easy" (FIG. 13-3). On the other side, you have the copilot thinking, "What a relief to not be flying as pilot in command today. Now I can relax and not worry about making any decisions." If such a crew isn't an accident waiting for a place to happen, I don't know what is.

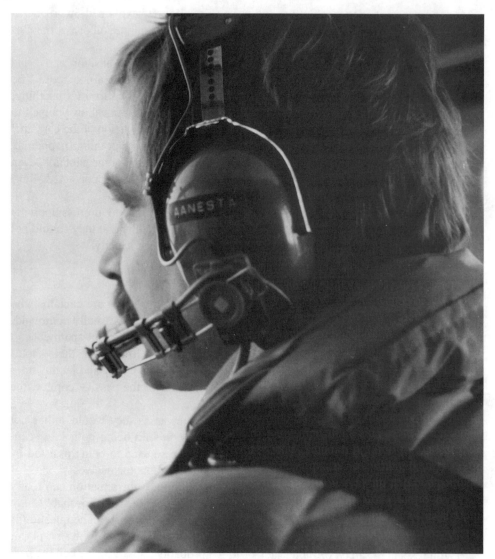

Fig. 13-3. *With an experienced captain flying as copilot, the pilot in command might lower his guard and not be as attentive as he would be with an inexperienced copilot.*

Being a good copilot, no matter what one's experience level, takes work. Personally, I've found I require a strong conscious effort to avoid the "he's-got-it" mind-set when I fly as copilot. It's easy to be an apathetic captain/copilot, but the easy way is not the right way.

262

Fig. 13-4. *Over-water hoist operations require the concerted effort of all crewmembers. The pilot in command needs the help of his copilot: Boeing Vertol CH-46 Sea Knight.*

WHAT TO DO

What's the solution to the problems of born-again copilots? Half the battle is recognizing the problems. Unfortunately, the captain-to-copilot transition is virtually ignored in most training programs. How to be a good copilot needs to be reiterated, not just to "standard" copilots, but to captains as well.

Born-again copilots must remember that being a good copilot means more than simply relieving the captain of purely mechanical tasks. He or she must be more than just an extra pair of eyes, hands, and feet—the captain needs the copilot's brain, too.

The captain, as pilot in command, still has the responsibility to make the decisions for the safe conduct of the flight, but the copilot should actively assist in the decision-making process. Perhaps more importantly, the copilot must function as a backup to the captain, keeping tabs on the whole situation, visually checking actions, mentally reviewing decisions, and verbally questioning anything he does not understand, thinks is being done improperly, or believes is unsafe (FIG. 13-4).

One thing is certain: Neither the passive nor the apathetic copilot has a place in the cockpit of any aircraft.

14
Ten Commandments

So Moses went down to the people and told them.

Exodus 19:25

BEFORE YOU HOVER OFF INTO THE SUNSET, I WANT TO LEAVE YOU WITH a list of rules to fly by. I'm not sure of the origin of "The Ten Commandments for Helicopter Flying" (FIG. 14-1), although I can assure you they weren't presented to humanity by Moses. I do know they were all over Fort Wolters, Texas, and Fort Rucker, Alabama, when I went through helicopter flight school with the Army. Copies of the commandments were printed on plaques, on prints suitable for framing, on beer mugs, on wallet-size cards, on poster-size posters, and many other items.

A lot of wives bought the commandments for their pilot-husbands as a graduation gift, perhaps in the hope that it would help prevent the wives from becoming widows. My wife, being of true Scottish blood, thought the plaques were too expensive. She made one instead, laboriously decoupaging one of the prints onto a piece of wood. It did come out rather nice and it hangs on a wall in my office.

The helicopter commandments are admittedly tongue-in-cheek, but they aren't fluff. There are a lot of ways to get yourself into trouble flying helicopters, but if you follow these commandments, you'll avoid most of them.

At least, they've worked for me.

The Ten Commandments for Helicopter Flying

**I. He who inspecteth not his aircraft gives his angels
cause to concern him.**
Inspect your aicraft carefully before each flight.

**II. Thou shalt not become airborne without first ascertaining
the level of thy propellant.**
It's better to spend minutes refueling than hours regretting.

**III. Let infinite discretion govern thy movement near the ground
for thy area of destruction is vast.**
Use extra caution while operating on or near the ground.

IV. Thy rotor rpm is thy staff of life, without it thou shall surely perish.
Low rpm is really dangerous. Keep it within the safe operating range.

**V. Thou shalt maintain thy speed between ten and four hundred feet
lest the earth rise and smite thee.**
Complete recovery is doubtful in case of power failure.

**VI. Thou shalt not make trial of thy center of gravity
lest thou dash thy foot against a stone.**
A few misplaced pounds may exceed the limits of your controls.

**VII. Thou shalt not let thy confidence exceed thy ability
for broad is the way to destruction.**
"I think I can make it" is on the list of famous last words.

**VIII. He that doeth his approach and alloweth the wind to turn behind
him shall surely make restitution.**
Make all approaches into the wind.

**IX. He who allows his tail rotor to catch in the thorns,
curseth his children and his children's children.**
Avoid tail low attitude while near the ground.

**X. Observe thou this parable lest on the morrow thy friends mourn
thee.**
Safety dwells with the safest man who flies his bird as safe he can.

Fig. 14-1. *The 10 Commandments for helicopter flying.*

I. He who inspecteth not his aircraft gives his angels cause to concern him.

For some reason, as most pilots gain experience and flight time, they become more and more blasé with their preflight inspections. Apparently, the "it-won't-happen-to-me" attitude becomes stronger the longer one flies and nothing does happen. All pilots, private or professional, male or female, young or old, are susceptible to this attitude.

The macho, devil-may-care image most pilots emulate practically demands that one take a relaxed attitude toward pre-flights. In reality, the quick walk-arounds of some older captains are not due so much to their competency ("Yes, ma'am, Old Joe's been flyin' so long he can spot an oil leak a hundert yards away with one eye closed!") as they are to plain old laziness and complacency ("Hey man, it worked yesterday and Bob the mechanic never misses a thing.")

But sometimes "Bob the mechanic," good as he is, does miss something, forgets something, or just plain screws up. I've found wrenches, rags, open panels, screwdrivers, unlocked fasteners, fuel caps missing or not properly attached, excessive oil on the fuselage, and many relatively minor problems (FIG. 14-2). The repair is usually simple. Most of the time "Bob" simply removes the "foreign object," closes the panel, fastens the fastener, finds another fuel cap, or wipes away the oil (spilled when refilling a reservoir), all the while with a rather sheepish smile on his face. But a few times the smile turns to a look of concern.

Fig. 14-2. *Check that the fuel caps are screwed on properly. The cap on this Enstrom F-28F has been painted with a white line; therefore, the pilot can quickly verify that the cap is in the correct position.*

I've heard of incidents where pilots found a hole in the tailboom from the previous flight and tail rotor blades mounted on backwards after an overhaul. Such things are obviously more serious. I know of one incident when the line mechanics and the pilots didn't notice that the star-shaped gust lock was still attached to the tail rotor of a Sikorsky S-61. When the pilots lifted the helicopter into a hover, they discovered they had very limited yaw control. Their immediate landing was just as hard as it was embarrassing.

It's not hard to embarrass yourself with a less than sterling preflight. Once, I was about to hit the starter of a Beechcraft Sundowner I had rented when the owner of the FBO suddenly appeared waving his arms for me to stop what I was doing. He made me get out of the airplane and pointed to the tow bar that was still fastened to the nose wheel. I forgot to remove the tow bar because I became preoccupied with brushing snow from the wings, per regulations, but there was no excuse for the oversight. I was lucky he still let me rent his airplane and even luckier he'd caught me before I started the engine.

When you preflight any aircraft after any maintenance work, visually check that everything is properly positioned and installed. A mechanic might ensure that everything has been properly handled when he signs off the logbooks; however, he might not be the last person to handle the aircraft before you climb aboard.

Wounded soldiers returning from Vietnam underwent three separate checks to ensure fitness for the long airplane trip home; consider the value of triple checking to ensure proper movement of control surfaces prior to take off. If you don't notice a problem before starting up, the problem might make its presence known to you in a very unpleasant way after you're in the air (FIG. 14-3).

What do you look for on a preflight inspection? Each helicopter is different and if you have a mechanic doing a daily inspection before you arrive, your inspection will be less detailed than if you have to do the inspection yourself.

Basically, look for the unusual: check the oil levels of the engine and gearboxes; look for unfastened fasteners and open inspection doors; remove all temporary covers; check the pitot tube and static ports for obstructions; drain some fuel from the sumps to check for water and look for excessive oil leaks; check fuel quantity, ideally visually checking the tanks, or checking the gauges; check the landing gear for condition and the tires for proper inflation (FIG. 14-4); check that the temperature indicated on temperature sensitive tapes is not too high; ensure that every nut and bolt that should be safety-wired is still safety-wired; check for bends, cracks, dents, and defects.

The best, most relied upon preflight inspection is the aptly-named walk-around. Start at one spot on the fuselage and walk around the machine, checking components as you go (FIG. 14-5). It doesn't matter if you start at the tail, or at the nose, or at the pilot's door. It doesn't matter if you walk around the craft clockwise or counterclockwise, do it the same way every time and do it with genuine interest. Do it as if, during the next hour or two, the condition of this machine will have a greater influence upon your life than nearly everything else in the world because it will.

Fig. 14-3. *Anyone that works around any aircraft might not want to admit to damage for fear of reprimand.*

Fig. 14-4. *Checking the S-61 landing gear.*

Bell Helicopter Textron

Fig. 14-5. *The walk-around inspection starts at one point on the fuselage and continues all the way around the aircraft: Bell 412.*

Your instructor will show you what to check. If he doesn't show you a good preflight, request it. If he still doesn't, get another instructor. After you have some time on the machine, ask him to go through a complete preflight again. You'll be surprised how much more you understand and therefore remember. When you check out in a different helicopter or airplane, be sure to go through a complete preflight with a pilot experienced in the machine.

Then it's up to you to be a professional about your preflight inspections every time, whether you fly for hire or for fun.

Don't give the angels cause to concern you. Inspect your aircraft carefully before each flight.

II. Thou shalt not become airborne without first ascertaining the level of thy propellant.

Checking how much fuel you have on board is an important part of a thorough preflight, so important that whoever thought of these commandments in the first place figured it should have its own commandment. Not a bad thought.

Aircraft engines have become extremely reliable since the early days of aviation and although things mechanical still sometimes go awry, the most common cause of engine failure is fuel starvation. Fuel starvation is usually the pilot's fault. Accident reports often call it "poor fuel management."

According to an old saying, "There are old pilots and there are bold pilots, but there aren't any old, bold pilots."

Old pilots know that Murphy's Law tends to make most flights longer than planned and that the best way to guard against the inevitable is to add fuel (FIG. 14-6). In the United States, you are allowed to fly a helicopter under visual flight rules with a fuel reserve of only 20 minutes. That is precious little fuel, believe me. Some pilots make it a habit to take on an additional 15 minutes of fuel for the wife and another 15 minutes for each kid. Others habitually round up all estimations when doing their fuel calculations. Whatever they do, the result is the same: an extra reserve above that required by the rules.

There are two basic ways to check your fuel and two ways to keep track of how much you have.

The first way to check the fuel is to physically look inside the tanks and confirm that, yes, there's fuel in them thar tanks. If the tanks are full to the top, this is a very good way to determine that you do, in fact, have full fuel. This is the accepted method for checking fuel on small helicopters, and even on the largest of machines, you can check fuel this way; however, if you have less than full fuel, figuring out how much you have becomes more difficult.

Fig. 14-6. *Be prepared for Murphy's Law and add extra fuel.*

Some tanks have visual fuel indicators with graduated scales on the sides that allow you to read directly how much fuel is in the tank (FIG. 14-7). This is a fairly accurate method as long as the machine is on level ground. If the helicopter is sitting on a slope, the gauge might indicate more or less than the actual amount.

Fig. 14-7. *The auxiliary fuel tank on this Bell 206B JetRanger has a sight gauge to visually check the fuel quantity.*

Other tanks have dipsticks that are usually fastened inside the tank or attached to the fuel cap. Some people make their own metal or wooden sticks for measuring the fuel quantity. The dipsticks are graduated and are fairly accurate as long as the ship is on level ground. A minor disadvantage is that it's sometimes hard to see the permanently mounted stick under some light conditions, which is why you might see a pilot carrying a flashlight while doing his preflight, even in bright sunlight.

Most large helicopters don't have visual indicators or dipsticks and the only way to check the fuel quantity, if it's not full, is to read the fuel gauges. Be sure to know what unit of measure the gauges on your machine read. Most helicopters manufactured in the United States show fuel in pounds, but some show it in gallons. European helicopters often measure fuel in pounds, in deference to the United States market, but might use liters or kilograms, too. Obviously, you'll be in for a big surprise if what you thought was 100 gallons of fuel turns out to be 100 liters—only one-fourth as much.

Contrary to popular belief, the indicator needles in fuel gauges don't always drop to zero when electrical power is turned off. In some aircraft, the needles freeze in position until the power is turned on again. (The fuel gauge in many cars does the same thing.) It's always prudent to switch on the power source to the gauges when checking them during preflight because you never know if someone refueled or defueled the machine or if fuel you left there the day before has leaked out, for example, because of a stuck drain valve.

What power source you switch on, depends on helicopter type. On most machines, switching on one or more of the batteries is sufficient since the fuel gauges run on DC power. If the gauges, or the indicating system, needs AC power, you'll have to switch on a ground inverter, too.

Do fuel gauges ever lie? You bet they do. Sometimes they fail outright and drop to zero. Then you're lucky, relatively speaking. At least a zero indication will catch your attention and alert you to the problem.

Unfortunately, most gauge failures are more insidious and harder to detect. A common failure is for the needle to freeze in some position while flying. Unless you're monitoring your fuel burn closely, it will probably be some time before you notice the gauge indicates there's more fuel on board than there should be. And it will go an even longer time before you figure out for sure that the gauge is showing too much and by then you really won't know how much you have.

Even sneakier are the gauges that decrease slightly or not at all and then suddenly jump down to what looks like the correct amount. Needles that stick (in all types of gauges and indicators) have plagued pilots for so long that the first action most pilots do when confronted with any odd indication is to tap the glass on the gauge. Of course, if your aircraft has EFIS displays instead of gauges you can bang them all you want and digital readouts will remain unchanged (until the screen breaks), but on the old electro-mechanical-type gauges you can often get a response.

Checking the fuel gauges is one way to determine your fuel load before flight and the main way to keep track of it while you're flying. The second way to keep track of your fuel is with your watch.

Using time to estimate fuel remaining is akin to dead reckoning navigation. It's sometimes imprecise, sometimes incredibly accurate, and always better than nothing at all.

Every airplane and every helicopter has a rule of thumb fuel burn. The rate of fuel consumption varies with power, temperature, altitude, humidity, fuel control adjustment, and a million other factors, but unless you're doing an awful lot of hovering or other high power maneuvers, your fuel consumption will pretty much average to within five to 10 percent of the rule of thumb. If you are doing a lot of hovering, you can always figure out a "fuel consumption while hovering" rule of thumb and use that instead of the standard "cruise flight" rule of thumb.

Always have a ballpark figure in your head when comparing time aloft and fuel remaining. It's a good double check of the gauges and a backup if you become suspicious of the fuel indication system.

For example, if you took off with three hours of fuel on board and, after one hour of flight, the fuel gauges still show three hours of fuel remaining, you can be fairly certain you have an indication failure. But if your gauges only show 45 minutes of fuel left, maybe it's the gauge acting up, or maybe it's because a fuel line has sprung a leak or the lineman didn't tighten the fuel cap down after refueling. The most prudent thing to do is land as soon as possible and check it out.

Remember to check your propellant during your preflight walk-around and always try to add more than you think you'll need. Often this isn't possible in a commercial operation when you must sacrifice some of your extra reserve in order to keep your payload as high as possible, but when you have room for more fuel, take it.

It's better to spend minutes refueling than hours regretting.

III. Let infinite discretion govern thy movement near the ground for thy area of destruction is vast.

Airplanes excel over other forms of transportation by virtue of their speed. Helicopters, to repeat the obvious, are different. The physical, aerodynamic laws governing all flight put a speed limit on helicopters at about 200 knots. As a consequence, conventional helicopters will never be the speed demons many of their fixed-wing cousins are.

What does all this have to do with movement close to the ground? Everything. Helicopters can't compete against airplanes in speed. If you have to go somewhere fast, you take an airplane. Because the air is thinner and offers less resistance at higher altitudes, greater speeds and better fuel efficiency are obtained at higher altitudes (up to a point); therefore, when you need to fly fast, you fly high. But many of the things you can do with aircraft require neither great speed nor great height; helicopters excel in these areas.

If you want to spot a robbery suspect making a getaway (FIG. 14-8), or take photographs of a disaster site, or transport an accident victim directly to a hospital, or airlift and set one hundred air conditioning units onto a roof, or inspect a powerline, you use a helicopter.

Statistics show that most airplane accidents occur during takeoff or landing, the time when airplanes are closest to the ground. The same is true for helicopters, but, by the nature of the jobs that are done with helicopters, they are close to the ground more often than airplanes. Many jobs helicopter pilots do place them constantly below altitudes that airplane pilots would consider safe. (See appendix D for tips on avoiding power lines.)

By the nature of the beast and the object of your mission as a helicopter pilot, you'll be exposing yourself to the danger of operating close to the ground much more often than if you were an airplane pilot (FIG. 14-9). That's just the way it is.

Extra vigilance is required. As a helicopter pilot, you'll have to be extremely aware of the space occupied by the helicopter, especially the rotors. This isn't easy. Except in the very small helicopters, you can't look out the back and see where the tail rotor is. If you plan to hover backward, you better be sure the area behind you is clear.

Bell Helicopter Textron

Fig. 14-8. *Flying a helicopter is the best way to follow a suspect fleeing the scene of a crime: Bell 206 JetRanger.*

Bell Helicopter Textron

Fig. 14-9. *Helicopters are low altitude vehicles—that's just the way it is: Bell 206 JetRanger.*

Without the advantage of rearward sight, it's impossible to imagine what's behind you. Even when you know the exact dimensions of the aircraft, how do you tell if there's five feet or 50 feet from the tail rotor to the concrete wall?

You can avoid the problem by not hovering backward, but hovering sideways instead. Or, if you can't turn to hover sideways, you can land where you are, walk back to where you want to go, and place a marker to tell you where to stop hovering backward. It might sound like a lot of bother just to check a few feet of space, but it will save you from a lot of embarrassment, or worse.

Care when flying close to the ground is for the benefit of other people, too. The main and tail rotors are obvious danger areas and the pilot must always ensure that people and objects are clear of his rotors.

The most dangerous time for other people is when the helicopter is on the ground with the rotors turning. The more hazardous of the two rotors is the tail rotor for three reasons. First, it spins so quickly that it is all but invisible; second, it's much closer to the ground and on small helicopters just about the right (or wrong) height for a person to walk into; and third, it's behind the pilot so there's a good chance he won't see someone walking toward it (FIG. 14-10).

This is not to say that the main rotors aren't dangerous, too, but most people do seem more aware of the big fan on top. Maybe it's from watching "M*A*S*H," but just about everyone who walks under a spinning main rotor bends down. This is not a bad procedure, but when the rotor system has reached normal operating rpm, it is usually unnecessary.

Fig. 14-10. *The protected fenestron-type tail rotor of the Aerospatiale SA-365N Dauphin is much safer than conventional tail rotors, but even it should be treated with respect.*

On the other hand, during rotor engagement and disengagement, nobody should stand under the rotor disc (FIG. 14-11). When the rotors are below normal operating rpm, they are more susceptible to the variance of the wind and are therefore not fully controllable by the pilot.

Fig. 14-11. *No one should be allowed underneath the rotor disc during rotor engagement and shutdown: Aerospatiale AS 332 Super Puma.*

The reason has to do with aerodynamics and physics: the faster the blades spin, the "stiffer" they become because of centrifugal force. To keep the blades from dipping too low many helicopters have droop stops that operate below a certain rotor rpm. Above the specified rpm during rotor engagement, the droop stops slip out of position so that the rotor blades can flap down to their limit, which might be a few feet above the ground, but ideally the pilot has control of the rotor disc by the time it spins up to "droop stop out rpm." Conversely, during disengagement the droop stops slip back in at a certain rpm to keep the blades from flapping down too low as they slow down.

Sudden gusts of wind will cause the blades to flap down. Because wind is unpredictable, and gusty winds are even more unpredictable, and because one never knows when a droop stop might not work properly, it's always a good idea to keep people outside the area directly below the rotor disc during rotor engagements and disengagements. It's also a good idea to keep your hands and feet on the controls until the rotors come to a complete stop. There have been incidents, and there will undoubtedly be more, of main rotor blades hitting the top of the cockpit, smacking into the ground, and even severing the entire tailboom during engagements and disengagements.

Another hazard of low-flying helicopters, often unrecognized by pilots as well as groundlings, is rotorwash. Rotorwash, like the wind, is invisible, but the effects are noticeable enough.

Small helicopters blow up quite a storm close to the ground. The big ones create minor hurricanes. The damage they cause can be considerable. For example, constant buffeting by Boeing 234 Chinook rotorwash over a period of five years caused structural damage to a terminal building at Norway's Forus Heliport.

Most people unfamiliar with helicopters don't realize how windy it can get when a helicopter lands, hovers, and takes off. I had to chuckle during a movie when one of the characters was waiting for someone to arrive by helicopter. He carefully combed his hair while the helicopter approached and then walked toward the landing spot. Seconds later the downwash hit him, not only mussing his hair, but covering him in dust as well.

When you're coming in to land and people are waiting for you, about the only thing you can do to warn them about the rotorwash is to stop in a high hover and descend slowly to the spot. This way the full force of the rotorwash won't hit all at once. Sometimes, conditions make this impossible and all you can do is hope that people have sense enough to cover there eyes or turn away. In the worst case, you might have to abandon the landing to avoid hurting someone.

Before taking off, you should take the time to warn anyone who will be standing nearby about rotorwash. They might not heed your advice, but at least you tried.

Finally, be aware of the possibilities of property damage by your helicopter's rotorwash when operating near the surface. Pay particular attention to lightweight objects, flimsy signs, and anything on wheels. I remember watching helplessly as the rotorwash of the helicopter I was taxiing caught an aluminum maintenance stand and pushed it across the ramp until it rolled into the fuselage of another helicopter. The mechanic who had last used the stand had apparently forgotten to set the wheel brakes.

Always watch what you do with a helicopter for your sake and the sake of others, and use extra caution while operating on or near the ground.

IV. Thy rotor rpm is thy staff of life, without it thou shall surely perish.

All of these commandments are vital and I would be hard pressed to list them in order of importance. But if I had to choose just one to place before all the others, the choice would be easy. It's this one.

If you remember nothing else from this book except this one commandment, you'll have at least learned the most important rule of helicopter flying (FIG. 14-12).

Why is proper rotor rpm so important? Recall that if the rotor isn't spinning fast enough, it can't make enough lift, nor can the tail rotor, and that's even worse because you'll lose yaw control even though the main rotor is still producing some lift. Because the tail rotor is spinning about five times faster than the main rotor and lift varies as the square of the velocity, the loss of yaw control might be more of a problem than the loss of lift from the main rotor.

If you let the rotor rpm increase too much, if it gets too high above the upper limit, there's no telling what might happen. Ideally, you'll be able to bring it back

Fig. 14-12. *Rotor rpm is your staff of life in any helicopter: Schweizer 300.*

down into limits and you'll only have to scrap the transmission, the rotor head, the tail rotor drive shaft and gearbox, the intermediate gearbox, and all the rotor blades.

Fortunately, maintaining proper rotor rpm during normal operations is a simple matter. With a reciprocating engine, you only need to monitor and adjust the throttle whenever you make collective changes. With turbine-powered helicopters, the fuel control unit will normally keep rotor rpm within limits for you.

During abnormal operations, remember the first step from the four-part helicopter emergency procedure: Fly the aircraft and maintain rotor rpm. Follow this procedure, discussed in chapter 10, and you will live to fly another day. You get the point.

Low rpm is very dangerous, likewise excessive rpm. Keep rpm within the safe operating range.

V. Thou shalt maintain thy speed between 10 and 400 feet lest the earth rise and smite thee.

This commandment refers to the height-velocity diagram (dead man's curve.) Every helicopter has such a curve (FIG. 14-13) and the exact figures vary by type, so consider 10 feet and 400 feet as rules of thumb only. On the other hand, as rules of thumb go, this one isn't too bad.

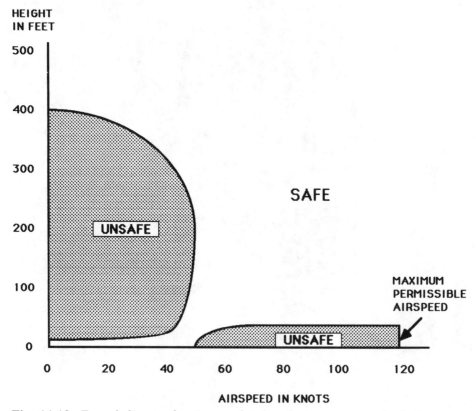

Fig. 14-13. *Every helicopter has its own height-velocity curve, so check the operating handbook or flight manual.*

Chapter 8 explains autorotations and the height-velocity diagram. If you are at all unsure why you should maintain airspeed between 10 and 400 feet, reread the section explaining the dead man's curve.

Remember, if you operate in the unsafe areas of the height-velocity diagram, complete recovery is doubtful in case of power failure.

VI. Thou shalt not make trial of thy center of gravity lest thou dash thy foot against a stone.

If there were ever a popularity contest for an aviation ground school subject, weight and balance would finish dead last, far behind everything else. I know of no other subject that consistently elicits so many moans and groans in the classroom and causes glazed-over eyeballs so quickly. This applies to student pilots and professional pilots.

Why this is, I'm not sure. I find weight and balance problems rather straightforward and simple, at the worst tedious. With an electronic calculator, most of the tediousness is eliminated. I suspect a lot of the subject's bad press has to do with the fact that a small error done early in the calculations can cause an answer that is not only out in left field, but totally out of the ballpark.

The best advice I can give to a new pilot who is struggling with weight and balance is to memorize the basic formula (moment = weight $\times$ arm), keep track of the labels that go along with the numbers (weight is in pounds, arm is in inches, weight times arm—moment—is in inch-pounds), and do the calculations carefully, writing everything. This will give you the right answer most of the time and help you find your mistakes the rest of the time.

Helicopters can be very sensitive to changes in center of gravity (CG). Fortunately, lateral (side-by-side) CG problems are less a concern in helicopters than in airplanes. The main reason is that airplanes carry fuel in their wings and the farther out on the wing you go, the longer the moment arm. If the fuel tanks in the right wing are full and the tanks in the left wing are empty, you would definitely notice it in an airplane.

Helicopters don't have wing tanks (obviously), although some have tanks along the sides of the fuselage or in sponsons over the wheels. These tanks are so close to the rotor mast that even if one is empty and the other is full, the change in the lateral center of gravity is barely noticeable.

Military helicopters often have wider sponsons to accommodate armament and external fuel tanks that make lateral CG more of a concern. For example, the USAF HH-3E flight manual carries this warning: "Asymmetric jettison of the external tanks during climb can result in rapid attainment of excessive roll rates and roll attitudes (20 degrees roll in 0.2 seconds)." But for the pilot of civilian helicopters, lateral CG is rarely a problem.

Longitudinal (front to rear) center of gravity is another matter. Virtually every helicopter can be loaded so that the CG ends up outside either the forward or aft limit, or both. The difference can be as small as an extra person in the front seat or a few bags or boxes in a cargo compartment (FIG. 14-14). In most cases, the problem won't be any more serious than a slightly more nose down attitude in forward flight or more nose up attitude in the flare. In the worst case, you might not even be able to lift into a hover.

Fig. 14-14. *Always check that cargo is loaded and secured properly. Check the center of gravity before takeoff.*

I know of one accident in an old Hiller 12E that was caused by improper loading of the small helicopter. With one person in the cockpit, the battery could be in either the front or the aft battery rack. With two people in the front, the battery was supposed to be in the aft rack.

The pilot picked up a passenger, but didn't move the battery to the aft rack. When he tried to take off, the machine nosed over so much the pilot couldn't move the cyclic far enough aft to prevent flying into the ground. Because the FAA determined that the cause was pilot error, the insurance company refused to pay up. The owner of the helicopter lost all he had.

Pilots often overlook CG changes due to fuel usage. This is particularly a problem in machines that have additional fuel tanks. Helicopter designers usually try to put the standard tanks as close as possible to the rotor mast because this causes the least CG change, but extra tanks often have to be placed farther and farther away from the mast. If the helicopter is loaded near the forward or aft limit when the tanks are full, improper fuel management in flight might cause the craft to go out of limits after a certain amount of fuel is burned off.

Last but not least, baggage compartments can be real culprits because they're often located in the aft part of the fuselage or even in the tailboom itself. A full load of baggage might be no problem if passengers are carried in the cabin, but if the passengers are let out at one place and the baggage is kept onboard to deliver to another spot, the aft CG limit could be exceeded. It's an unnerving experience to lift up into a hover and find that full forward cyclic is insufficient to keep the helicopter from nosing up and moving backwards (FIG. 14-15).

So, as boring as the subject might be, learn and abide by the weight and balance limits of your helicopter. A few misplaced pounds might exceed the limits of your controls.

United Technologies Sikorsky Aircraft

Fig. 14-15. *External loads are normally carried directly below the rotor mast, which is a good CG position: Sikorsky UH-60 Black Hawk.*

VII. Thou shalt not let thy confidence exceed thy ability, for broad is the way to destruction.

Ah, confidence. If only those that have too little could get some from those that have too much. Psychologists say lack of confidence is one of the great inhibitors to success in this world while accident investigators know, without a doubt, that too much confidence easily leads to destruction. The line between the two is thin, indeed.

Years ago, when I was a gung-ho Jolly Green Giant pilot in the Air Force, I was talking with one of the HC-130 Hercules pilots in our squadron. He told me his personal credo for staying out of trouble: "I try not to get too creative in the cockpit." I've always remembered that.

There are ways to get the job done and there are ways to get the job done. When flying difficult rescue missions in the harsh environment of the North Atlantic, as we were at the time, it's easy to find ways to cut corners and bend regulations (FIG. 14-16). The Air Force way was often very time-consuming, but it did get the job done, most of the time. Usually, there was no need to get creative in the cockpit.

Fig. 14-16. *It's tempting to cut corners when the mission is important, but it's not healthy to get too creative in the cockpit: Lockheed HC-130 refueling Sikorsky CH-53E Super Stallions.*

If there was the slightest hint of urgency in the mission, a lot of Air Force pilots succumbed to the macho "regulations-be-damned-let's-get-the-job-done" attitude. I had to grudgingly admire this pilot's much more mature attitude. His confidence level was in perfect balance between too much and too little.

It's been said that the young pilot's enemy is inexperience and the old pilot's enemy is complacency. Overconfidence, however, can hit every pilot, at any experience level. It's just a matter of degree.

If you're good at doing something, touch-and-goes, hovering autorotations, confined area landings, and the like, you can become over-confident about it, even if you don't have much flight time. High-time pilots are just good at a lot more things than low-time pilots; therefore, there's a lot more they can become overconfident about.

On the other hand, low-time pilots seem more prone to become puffed up with their own, albeit limited, abilities. Good high-time pilots usually know their limits much better than low-time pilots. If they're really good, they not only know what they're good at but also what they tend to get lazy about. They know the situations that are conducive to mistakes and force themselves to continually check and double-check their own actions. This is what distinguishes the great pilots from the good ones.

A Trump Air pilot I used to fly with had an effective technique to catch his and his copilot's mistakes. Several times during a flight, when there were no pressing tasks at hand, he'd survey the instruments and switches before us and ask, "What's wrong with this picture?" Then we'd both look to see if there was a switch that was in the wrong position, or a radio or navaid that should be tuned to another frequency, or an instrument that was giving an unusual indication. Sometimes we'd find something minor amiss, most of the time we didn't. But at least we looked.

We didn't do anything more than what we should have been doing all the time, but he had enough experience to know he sometimes had to give himself, and other pilots, a reminder.

One of the most difficult things about flying is knowing your own limits and the limits of your aircraft. Strictly speaking, the only way you can really know a limit is by exceeding it; then you back off by one notch. That's not a wise way to find your limits unless you happen to be training in a simulator. Of course, the manufacturer has very kindly given us the aircraft's limits for a number of things and it is imperative that you do not exceed these on purpose (FIG. 14-17).

Unfortunately, some limits are rather vague and left up to the judgment and interpretation of the pilot. How do you know when you've reached these limits? To be honest, you don't know, unless something breaks and then you know you've both reached and exceeded a limit or two. The only sensible thing to do is play it extra safe by staying well inside the flight manual limitations for your aircraft.

Your own ability is even harder to gauge than the capability of the aircraft because it is constantly changing. As you gain hours and experience, your ability in a particular aircraft increases. Switch to another aircraft and, although you won't be back to square one, you will have a new learning curve to climb.

Lay off flying for a few days and your skill level deteriorates a tad. Lay off a few weeks and you'll really feel rusty. Lay off a few months and you'll be embarrassed by how much you have forgotten.

Fly several hours every day for two weeks in a row and your skill level will be way up, but so will your fatigue level; at some point, fatigue will cause your skill level to drop. I almost had to fail one of the best pilots I have ever known on a company checkride because he was so fatigued from working extra days that his judgment and skill level were way down. He even looked like hell, he was so tired. I passed him only because I knew from other checkrides he could have done much better if he were rested and because he was scheduled to go home for three weeks as soon as we were done. Failing him would only have delayed the remedy to the problem.

Your own skill level will change during a long flight as your body tires and your attention level wanes. Physiological factors such as blood sugar level and sleep (or lack of it) also play a part. When you think about it, it's amazing any of us ever fly safely. The only way to stay safe is to put extra limits around your limits and cushions around your capabilities. Don't fly consistently to the edge of your capability, but rather inside a more limited regime. Then if you ever need that extra something to pull yourself out of a hairy situation, you'll have it.

McDonnell Douglas Helicopter Corporation

Fig. 14-17. *Aerobatic maneuvers exceed the limits of most helicopters and are therefore forbidden. The McDonnell Douglas AH-64A Apache is an exception.*

"Know thyself," is one of the inscriptions at the Delphic Oracle in Ancient Greece, attributed to the Seven Sages, approximately 600 B.C. It's a good axiom for pilots in the 20th Century. "I think I can make it" is a bad axiom.

VIII. He that doeth his approach and alloweth the wind to turn behind him shall surely make restitution.

Airplane pilots have it easy because they take off and land from nice, long, smooth runways. An airport with multiple runways means that a pilot can usually take off within 45 degrees of the wind. If a pilot must take off downwind, he can open up a book and find out how long the takeoff run is going to be and compare that to the runway length. He can call up the tower and find out what the wind is to the knot and degree. If the airport doesn't have a tower, he can look at the wind sock.

Helicopter pilots aren't so lucky. Well, all right, at an airport we get all the same information airplane pilots get and thereafter we usually can take off directly into the wind regardless of the runway direction. But that's only at airports.

Helicopters are the ultimate off-road vehicles and there are few places in the boondocks that have 24-hour ATIS broadcasts. How to find the wind direction and estimate its speed is something every helicopter pilot must learn. Taking off and landing downwind is asking for trouble (FIG. 14-18).

Enstrom Helicopter Corporation

Fig. 14-18. *As a helicopter pilot, you must always maintain an awareness of the wind direction: Enstrom TH28.*

The section regarding wind in chapter 7 explains in detail the hazards of taking off and landing downwind or crosswind. Go back and read it again, if you've forgotten.

Sometimes you'll have no choice and won't be able to take off directly into the wind. If you know the potential problems involved when you do this, you can prepare yourself to meet the challenge. But the general rule is, if you have a choice, make all your approaches and takeoffs into the wind.

IX. He who allows his tail rotor to catch in the thorns, curseth his children and his children's children.

Making an approach and landing, or trying to land, downwind is one good way to catch your tail rotor in the thorns. Another way is to be inattentive while hovering in a confined area (FIG. 14-19). You can also do it by applying too much aft cyclic while in a low hover or by allowing the aircraft to descend while hovering backwards. A center of gravity near the aft limit (or beyond) will give the helicopter a tail-low attitude and increase the chance of the tail rotor hitting the ground.

McDonnell Douglas Helicopter Company

Fig. 14-19. *Be acutely aware of the tail rotor's position and its relationship to obstacles: McDonnell Douglas MD 500.*

Why is this bad? Besides putting a groove in the runway where there shouldn't be one, it could also cause you serious control problems. Unless you just kiss the ground with the tips of the blades, a tail strike will probably ruin your whole day, and a lot more days to come.

Tail rotors are incredibly fine tuned and balanced and the loss of only ounces from just one blade can be enough to create an imbalance and horrendous vibrations. Some, if not all, yaw control will be lost. If you don't lose all of it right away, there's a good chance you will after a matter of seconds if the vibrations are severe enough. The tail rotor gearbox, intermediate gearbox (if there is one), and tail rotor shaft are built to withstand a lot of punishment, but not that induced by a damaged tail rotor. The machine will literally shake itself to pieces.

This commandment is next to last and there isn't much left to say about it, nevertheless it's still vitally important. Avoid a tail low attitude while near the ground.

X. Observe thou this parable lest on the morrow thy friends mourn thee.

Can I make you a safe pilot? No.

Can your instructor make you a safe pilot? No.

Can the FAA make you a safe pilot? Not really. They can make you a safe former pilot by revoking your certificate if you are a chronically unsafe operator, but they can't really make you a safe pilot.

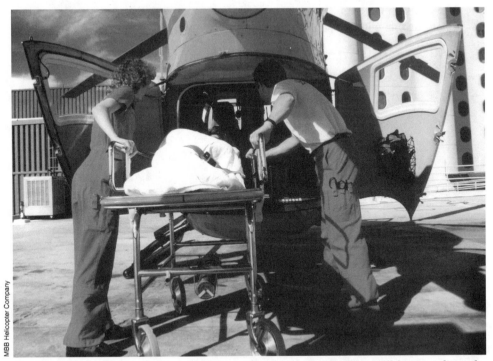

Fig. 14-20. *What kind of helicopter pilot will you be? Obviously, you'll be a credit to the profession and perhaps fly the MBB BK 117.*

MBB Helicopter Company

The only person that can make you a safe pilot is you. You have to have the desire, the knowledge, the maturity, and, sometimes, the courage to be a safe pilot. You have to make the decision yourself. It's completely on your shoulders.

I'm reminded of the old joke about how many psychiatrists it takes to change a light bulb: "Only one, but the light bulb has to want to change."

The joke is funny because the inference is true, at least in its allusion to psychiatry and human nature. A psychiatrist can't make your mental health better. He can only help you find out what's bugging you and show you ways and techniques to point you in the right direction. If you want to change, you have to do that yourself. The good news is that most psychiatrists and psychologists believe you really can control your behavior, despite your past (contrary to what was generally believed by Freud and his early followers).

Chuck Yeager, retired Air Force General and the first person to fly faster than the speed of sound, puts it this way, "If you want to grow old as a pilot, you've got to know when to push it, and when to back off."

So it's up to you. You have to make the choice. Are you going to be a safe pilot and a credit to yourself, your employer, and the profession (FIG. 14-20), or are you going to be the opposite?

If you follow these commandments religiously, you will be a safe pilot. If you don't, sooner or later you'll give your friends and family cause to mourn you.

The decision is yours.

Safety dwells with the safest man who flies his aircraft as safe as he can.

Postflight

Whatever you can do, or dream you can, begin it. Boldness has genius, power, and magic in it.

Johann Wolfgang von Goethe

WE'VE COME A LONG WAY, YOU AND I. WE STARTED OUT WITH FIVE myths and ended up with 10 commandments. In between, we've covered a lot of ground and about 20 years and 9,000 hours worth of personal flight experience. I sincerely hope you've enjoyed reading this book as much as I've enjoyed writing it. Actually, I hope you've enjoyed reading it a lot more than I've enjoyed writing it. Writing *Learning to Fly Helicopters* has been very hard work. But it's been worth it.

If you have any questions or comments, please feel free to write to me in care of TAB Books, a division of McGraw-Hill. I'll do my best to reply to you as quickly as I can.

I also encourage you to read other books about helicopters: how to fly them, what they do, and their history. I think you'll find the history of helicopters particularly fascinating.

One final word. Learning to fly helicopters isn't easy. But don't let that scare you. If flying helicopters is what you want to do, then begin and don't give up. You're going to get discouraged sometimes. Hang in there. Keep practicing and it will come. Remember, "Persistence and determination alone are omnipotent." Before you know it, you'll get the feeling, too.

Fly safe and watch your rotor rpm!

Appendix A
Common Civilian Helicopters Flying Today

DECIDING WHICH HELICOPTERS TO INCLUDE WAS THE MOST DIFFICULT thing about writing this section. Including every helicopter ever built would have been a monumental task, far beyond the scope of this book. What about "helicopters presently being manufactured?" Not a bad idea, but too limited. There are many helicopters still in service that aren't being manufactured now. To leave them out would have been a disservice to many readers who might find themselves flying these older machines. Helicopters, like all aircraft, stay around a long time.

The ideal solution was "Common Civilian Helicopters Flying Today." Which helicopters are "common" and which are "uncommon" can only be an arbitrary decision. Numerically, some machines are a lot more common than others, but many of these are older helicopters slowly aging toward retirement. On the other hand, some of the newer machines aren't even in production yet, but have a great potential for the future; thus, I had to make some choices.

I decided to limit this section to civilian helicopters because this book is primarily for civilian pilots; however, because most civilian helicopters have military counterparts, you can get a good idea what many military machines are like. In addition, there are photographs of many military helicopters throughout the book.

One thing you'll quickly learn in the helicopter business is that it's always in a state of flux. Many of the older helicopter companies have been bought up by other companies. For example, the Hughes 300 is now manufactured by the Schweizer Aircraft Corporation and the Hughes 500 is made by the McDonnell Douglas Helicopter Company. The old Hiller aircraft have found new life with the Rogerson Hiller Corporation (not to be confused with the Robinson Helicopter Company).

Some companies cooperate for a while to produce one helicopter type, others merge permanently. Westland and Agusta are jointly producing the E.H. Industries EH 101, for example. MBB (Messerschmitt-Bölkow-Blohm) makes the Bölkow BO-105 and the Bölkow-Kawasaki BK 117. The relationship between Kawasaki, Messerschmitt, Bölkow, and Blohm, was always difficult to understand but it doesn't really matter now because they've merged with Aerospatiale to form European Helicopters. Aerospatiale is actually the result of numerous mergers and acquisitions in the French aviation industry.

To expand international markets, many companies allow other companies to manufacture their machines under license; thus, Westland and Agusta have produced Sikorsky aircraft; Agusta also builds Bell helicopters. As a result, someday you might fly Bell a 212 that is an AB (Agusta Bell) 212.

Cooperation among helicopter companies will only increase in the future because the cost of developing new aircraft is so high. The United States Army's light helicopter experimental (LHX) program ended up being a competition between two sets of manufacturers: a Bell-McDonnell Douglas team and a Sikorsky-Boeing team. (The Sikorsky-Boeing team won the contract in 1991.) On the other hand, Bell and Boeing are working together on the V-22 tilt-rotor program.

Calling a helicopter by the wrong name would be a big problem if helicopter people weren't so tolerant. But they are, and nobody will get bent out of shape if you tell them you just flew a Hughes 300 when it really was a Schweizer 300. You'll soon learn how to tell the two apart. (The only sure way is to look for the manufacturer's nameplate.) Of course, by the time you can tell a Hughes 300 from a Schweizer 300 some other company might be making the 300-series.

HELICOPTERS TODAY

I have tried to provide the most accurate information possible for the helicopters shown in this section; however, because there are often numerous variants of many of the machines, there are undoubtedly some discrepancies. I apologize if any of the aircraft are incorrectly represented.

With respect to the data: the seating capacity is the total number of seats available for pilot(s) and passengers; speed is the fast cruise airspeed listed by the manufacturer; and the range is the maximum distance that can be flown with a full load at the maximum takeoff weight quoted.

Aerospatiale SA 315B Lama
Engines: single turbine
Seats: 4—5
Speed: 100 knots
Range: 280 nautical miles
Weight: 4,300 pounds
Rotor diameter: 36 feet

Aerospatiale Helicopter Corporation

Aerospatiale SA 316B Alouette III
Engines: single turbine
Seats: 4–5
Speed: 140 knots
Range: 390 nautical miles
Weight: 3,980 pounds
Rotor diameter: 34.5 feet

Aerospatiale Helicopter Corporation

Aerospatiale AS 350B Ecureuil
Engines: single turbine
Seats: 6–7
Speed: 125 knots
Range: 400 nautical miles
Weight: 4,300 pounds
Rotor diameter: 35 feet

Aerospatiale Helicopter Corporation

Aerospatiale AS 355F-2 Twin Star
Engines: twin turbine
Seats: 6–7
Speed: 122 knots
Range: 382 nautical miles
Weight: 5,600 pounds
Rotor diameter: 35 feet

Aerospatiale Helicopter Corporation

Aerospatiale Helicopter Corporation

Aerospatiale SA-365N Dauphin 2
Engines: twin turbine
Seats: 13–14
Speed: 150 knots
Range: 475 nautical miles
Weight: 9,369 pounds
Rotor diameter: 39 feet

Aerospatiale AS 332L Super Puma
Engines: twin turbine
Seats: 20–26
Speed: 144 knots
Range: 464 nautical miles
Weight: 18,960 pounds
Rotor diameter: 51 feet

Agusta Aerospace Corporation

Agusta 109C
Engines: twin turbine
Seats: 7–8
Speed: 150 knots
Range: 420 nautical miles
Weight: 5,997 pounds
Rotor diameter: 36 feet

Bell 47
Engines: single piston
Seats: 2−3
Speed: 75 knots
Weight: 2,950 pounds
Rotor diameter: 37 feet

Bell Helicopter Textron

Bell 206B III JetRanger
Engines: single turbine
Seats: 4−5
Speed: 132 knots
Range: 400 nautical miles
Weight: 3,200 pounds
Rotor diameter: 33 feet

Bell Helicopter Textron

Bell 206L LongRanger
Engines: single turbine
Seats: 6−7
Speed: 130 knots
Range: 360 nautical miles
Weight: 4,150 pounds
Rotor diameter: 37 feet

Bell Helicopter Textron

Bell 222B
Engines: twin turbine
Seats: 8−10
Speed: 145 knots
Range: 400 nautical miles
Weight: 8,270 pounds
Rotor diameter: 42 feet

Bell Helicopter Textron

Bell 212

Engines: twin turbine
Seats: 13 – 15
Speed: 127 knots
Range: 260 nautical miles
Weight: 11,200 pounds
Rotor diameter: 48 feet

Bell 412HP

Engines: twin turbine
Seats: 11 – 15
Speed: 144 knots
Range: 400 nautical miles
Weight: 11,900 pounds
Rotor diameter: 46 feet

Bell 214ST

Engines: twin turbine
Seats: 17 – 20
Speed: 161 knots
Range: 465 nautical miles
Weight: 17,500 pounds
Rotor diameter: 52 feet

299

Boeing 234
Engines: twin turbine
Seats: 44 – 48
Speed: 145 knots
Range: 500 nautical miles
Weight: 48,000 pounds
Rotor diameter: 60 feet

E.H. Industries EH 101
Engines: three turbine
Seats: 32
Speed: 150 knots
Range: 500 nautical miles
Weight: 19,695 pounds
Rotor diameter: 61 feet

Enstrom F-28F Falcon
Engines: single piston
Seats: 2 – 3
Speed: 112 knots
Range: 228 nautical miles
Weight: 2,600 pounds
Rotor diameter: 32 feet

Enstrom F280FX Shark
Engines: single piston
Seats: 2 – 3
Speed: 117 knots
Range: 260 nautical miles
Weight: 2,600 pounds
Rotor diameter: 32 feet

MBB BO-105 LS A-3
Engines: twin turbine
Seats: 5 – 6
Speed: 129 knots
Range: 278 nautical miles
Weight: 5,732 pounds
Rotor diameter: 32 feet

MBB BK-117 B-1
Engines: twin turbine
Seats: 8 – 11
Speed: 134 knots
Range: 297 nautical miles
Weight: 7,055 pounds
Rotor diameter: 36 feet

301

McDonnell Douglas MD 500 E
Engines: single turbine
Seats: 4–5
Speed: 145 knots
Range: 285 nautical miles
Weight: 3,550 pounds
Rotor diameter: 26.4 feet

McDonnell Douglas MD 520N
Engines: single turbine
Seats: 5
Speed: 134 knots
Range: 217 nautical miles
Weight: 3,350 pounds
Rotor diameter: 26.4 feet
(No tail rotor.)

Robinson R22 Beta
Engines: single piston
Seats: 2
Speed: 110 knots
Range: 300 nautical miles
Weight: 1,370 pounds
Rotor diameter: 25 feet

Robinson Helicopter Company

Robinson R44
Engines: single piston
Seats: 2
Speed: 113 knots
Range: 400 nautical miles
Weight: 2,400 pounds
Rotor diameter: 33 feet

Rogerson Hiller Corporation

Rogerson Hiller UH-12E
Engines: single piston
Seats: 2−3
Speed: 90 knots
Range: 280 nautical miles
Weight: 3,100 pounds
Rotor diameter: 35 feet

Schweizer Aircraft Corporation

Schweizer 300
Engines: single piston
Seats: 2−3
Speed: 75 knots
Range: 220 nautical miles
Weight: 1,905 pounds
Rotor diameter: 25 feet

Sikorsky Aircraft S-58T
Engines: twin turbine
Seats: 16
Speed: 110 knots
Range: 242 nautical miles
Weight: 13,000 pounds
Rotor diameter: 56 feet

Sikorsky S-61N
Engines: twin turbine
Seats: 26–30
Speed: 121 knots
Range: 430 nautical miles
Weight: 20,500 pounds
Rotor diameter: 62 feet

Sikorsky S-76 Mk 2
Engines: twin turbine
Seats: 12–14
Speed: 145 knots
Range: 400 knots
Weight: 10,325 pounds
Rotor diameter: 44 feet

United Technologies Sikorsky Aircraft

Appendix B
Helicopter Flight Test Standards

The Aviations Standards National Field Office of the FAA has developed the Private Pilot Practical Test Standards book (FAA-S-8081-1A) as the standard to be used by FAA inspectors and designated pilot examiners when conducting airman practical tests (oral and flight tests). Flight instructors are expected to use the book when preparing applicants for practical tests.

The book contains the private pilot practical test standards for airplanes (single- and multiengine, land and sea), rotorcraft (helicopter and gyroplane), glider, and lighter-than-air (airship and free-balloon). Only the section pertaining to helicopters is included in this appendix. The complete book may be obtained from the Superintendent of Documents, U.S. Government Printing Office, Washington, DC 20402.

IMPORTANT TO KNOW

Applicants should know some important points regarding the standards.

First, it is the responsibility of your flight instructor to train you to the acceptable standards as outlined in the objective of each task within the appropriate practical test standard. Your flight instructor must certify that you are able to perform safely as a private pilot and are competent to pass the required practical test for the certificate or rating sought.

Second, the flight examiner (whether an FAA inspector or an FAA designated pilot examiner) is responsible for determining that you meet the standards outline in the objective of each task within the appropriate practical test standard. For each task that involves "knowledge only" elements, the examiner will orally quiz you. For each task that involves both "knowledge and skill" elements, the examiner will orally quiz you regarding the knowledge elements and ask you to perform the skill elements. Oral questioning might be used at any time during the practical test.

Third, the examiner is not required to follow the precise order in which the tasks appear in each section and may combine tasks with similar objectives to save time. The examiner may also, for a valid reason, decide to evaluate certain tasks orally. Valid reasons are that the tasks do not conform to the manufacturer's recommendations or operating limitations or that are impractical at the time of the test: night flying, operations over congested areas, unsuitable terrain, and the like.

Fourth, the examiner is permitted and will probably provide realistic distractions throughout the practical test. These distractions might be used to evaluate your ability to divide your attention while maintaining safe flight and include simulating engine failure, identifying features on the ground, removing objects from the glove compartment or map case, questioning.

Fifth, special emphasis will be placed on areas of aircraft operation which are most critical to flight safety. Among these are correct aircraft control and sound judgment in decision making. The examiner will also emphasize spatial disorientation, collision avoidance, wake turbulence avoidance, low-level wind shear, and proper use of checklist.

PREREQUISITES

Before you take the private pilot practical test, you are required to:

- Pass the appropriate pilot written test since the beginning of the 24th month before the month in which the flight test is taken.

- Obtain the applicable instruction and aeronautical experience prescribed for the pilot certificate or rating sought.

- Possess a current medical certificate appropriate to the certificate or rating sought.

- Meet the age requirement for the issuance of the certificate or rating sought.

- Obtain a written statement from an appropriately certificated flight instructor certifying that you have been given flight instruction in preparation for the practical test within 60 days preceding the date of applications. The statement shall also state that the instructor finds you competent to pass the practical test, and that the applicant has a satisfactory knowledge of the subject areas in which a deficiency was indicated by the airman written test report.

AIRCRAFT AND EQUIPMENT REQUIREMENTS

You are required to provide an appropriate and airworthy aircraft for the practical test. The aircraft must be equipped for, and its operating limitations must not prohibit the pilot operations required on the test.

SATISFACTORY PERFORMANCE

Your ability to perform the required tasks is based on:

- Executing the tasks within the aircraft's performance capabilities and limitations, including use of the aircraft's systems.

- Executing emergency procedures and maneuvers appropriate to the aircraft.

- Piloting the aircraft with smoothness and accuracy.

- Exercising good judgment.

- Applying aeronautical knowledge.

- Showing mastery of the aircraft within the standards outlined in the test standards, with the successful outcome of a task never seriously in doubt.

UNSATISFACTORY PERFORMANCE

If, in the judgment of the examiner, you do not meet the standards of performance of any task performed, the associated pilot operation is failed and therefore, the practical test is failed.

You or the examiner may discontinue the test at any time after the failure of a pilot operation. The test will be continued only with your consent. If discontinued however, you will only get credit for those tasks satisfactorily performed, although during the retest, any task might be reevaluated at the discretion of the examiner, including those previously passed.

1. PREFLIGHT PREPARATION

a. Task: Certificates and Documents

References: FAR Parts 61 and 91, AC 61-23, Helicopter Flight Manual (see note at the end of this appendix).

Objective: To determine that the applicant:

1) Exhibits knowledge by explaining the appropriate—
   (a) pilot certificate, privileges and limitations.
   (b) medical certificate, class and duration.
   (c) personal pilot logbook or flight record.
   (d) FCC station license and operator's permit, if applicable.
2) Exhibits knowledge by locating and explaining the significance and impor-

tance of the—
(a) airworthiness and registration certificates.
(b) operating limitations, handbooks, or manuals.
(c) equipment list.
(d) weight-and-balance data.
(e) maintenance requirements and appropriate records.

b. Task: Obtaining Weather Information

References: AC 00-6, ac 00-45, AC61-23, AC 61-84.

Objective: To determine that the applicant:

1) Exhibits knowledge of aviation weather information by obtaining, reading, and analyzing—
(a) weather reports and forecasts.
(b) weather charts.
(c) pilot weather reports.
(d) SIGMETS AND AIRMETS.
(e) Notices to Airmen.
(f) wind-shear reports.
2) Makes a competent go/no-go decision based on the available weather information.

c. Task: Determining Performance and Limitations

References: AC 61-21, AC 61-23, AC 61-84, Helicopter Flight Manual.

Objective: To determine that the applicant:

1) Exhibits knowledge by explaining helicopter weight and balance, performance, and limitations, including adverse aerodynamic effects of exceeding the limits.
2) Uses available and appropriate performance charts, tables, and data.
3) Computes weight and balance, and determines if weight and center of gravity will be within limits during all phases of flight.
4) Calculates helicopter performance, considering density altitude, wind, terrain, and other pertinent conditions.
5) Describes the effects of atmospheric conditions on helicopter performance.
6) Makes a competent decision on whether the required performance is within operating limitations of the helicopter.

d. Task: Cross-Country Flight Planning

References: AC 61-21, AC 61-23, AC 61-84.

Objective: To determine that the applicant:

1) Exhibits knowledge by planning, within 30 minutes, a VFR slope-country flight of a duration near the range of the helicopter, considering fuel and loading.
2) Selects and uses current and appropriate aeronautical charts.
3) Plots a course for the intended route of flight with fuel stops, as necessary.
4) Selects prominent en route checkpoints.
5) Computes flight time, headings, and fuel requirements.
6) Selects appropriate radio navigation aids and communications facilities.
7) Identifies airspace, obstructions, and alternate heliports/airports.
8) Extracts pertinent information from the Airport/Facility Directory and other flight publications, including NOTAMs.
9) Completes a navigation log.
10) Completes and files a VFR flight plan.

e. Task: Operation of Helicopter Systems

References: AC 61-13, Helicopter Flight Manual.

Objective: To determine that the applicant exhibits knowledge by explaining the helicopter systems and operation, including, as appropriate:

1) Primary flight controls, trim, and stability systems.
2) Pitot static system and associated flight instruments.
3) Landing gear.
4) Rotor system.
5) Powerplant.
6) Fuel system.
7) Oil system (engine and transmission).
8) Hydraulic system.
9) Electrical system.
10) Environmental system.
11) Deice and anti-ice systems.
12) Avionics.

f. Task: Aeromedical Factors

References: AC 61-21, AC 67-2, AIM.

Objective: To determine that the applicant:

1) Exhibits knowledge of the elements related to aeromedical factors, including the symptoms, effects, and corrective action of
   (a) hypoxia.
   (b) hyperventilation.
   (c) middle ear and sinus problems.

(d) spatial disorientation.

(e) motion sickness.

(f) carbon monoxide poisoning.

2) Exhibits knowledge of the effects of alcohol and drugs, and the relationship to flight safety.

3) Exhibits knowledge of nitrogen excesses during scuba dives, and how this affects a pilot or passenger during flight.

2. GROUND OPERATIONS

a. Task: Visual Inspection

References: AC 61-13, Helicopter Flight Manual.

Objective: To determine that the applicant:

1) Exhibits knowledge of helicopter visual inspection by explaining the reasons for checking all items.

2) Inspects the helicopter by following a checklist.

3) Determines that the helicopter is in condition for safe flight, emphasizing—

(a) fuel quantity, grade, and type.

(b) fuel contamination safeguards.

(c) oil quantity, grade, and type.

(d) fuel, oil, and hydraulic leaks.

(e) flight controls.

(f) structural damage.

(g) rotor blade tiedown and removal.

(h) blade damper positioning, where appropriate.

(i) rotor blade positioning for starting, or clutch engagement, as necessary.

(j) strut extension, as appropriate.

(k) ice and frost removal.

(l) security of baggage, cargo, and equipment.

b. Task: Cockpit Management

References: AC 61-13.

Objective: To determine that the applicant:

1) Exhibits knowledge of cockpit management by explaining related safety and efficiency factors.

2) Organizes and arranges the material and equipment in an efficient manner.

3) Ensures that the safety belts and shoulder harnesses are fastened.

4) Adjusts and locks the anti-torque pedals and pilot's seat to a safe position and ensures full control movement.

5) Briefs occupants on the use of safety belts and emergency procedures, including, the use of flotation gear and pyrotechnic signalling devices, when aboard.

c. Task: Starting Engine

References: AC 61-21, AC 91-13, AC 91-55, Helicopter Flight Manual.

Objective: To determine that the applicant:

1) Exhibits knowledge by explaining engine starting procedures, including starting under various atmospheric conditions.
2) Performs all the items on the checklist.
3) Accomplishes correct starting procedures with emphasis on—
   (a) positioning the helicopter to avoid creating hazards.
   (b) determining the area is clear.
   (c) adjusting the flight controls and friction settings.
   (d) setting the brakes, if so equipped.
   (e) preventing helicopter movement after engine start.
   (f) avoiding hot starts and engine overspeed, and observing cooling limitations.
   (g) safety precautions related to engine start/rotor engagement.
   (h) checking engine instruments and caution/warning systems.

d. Task: Pretakeoff Check

References: AC 61-13, Helicopter Flight Manual.

Objective: To determine that the applicant:

1) Exhibits knowledge of the pretakeoff check by explaining the reasons for checking all items.
2) Positions the helicopter to avoid creating hazards.
3) Divides attention inside and outside of the cockpit.
4) Accomplishes the checklist items.
5) Ensures that the helicopter is in safe operating condition.
6) Reviews takeoff performance and emergency procedures.
7) Obtains and interprets takeoff and departure clearances.

e. Task: Postflight Procedures

References: AC 61-13, Helicopter Flight Manual.

Objective: To determine that the applicant:

1) Exhibits knowledge by explaining the elements of postflight procedures, including parking, cool-down and temperature stabilization, shutdown, securing, and postflight inspection.

2) Selects the designated or suitable parking area, considering wind conditions and obstructions.
3) Parks the helicopter properly.
4) Follows the recommended procedure for engine shutdown, cockpit securing, and disembarking passengers.
5) Secures the helicopter properly.
6) Performs a satisfactory postflight inspection.

3. HOVERING AND MANEUVERING BY GROUND REFERENCES

a. Task: Vertical Takeoff

References: AC 61-13, Helicopter Flight Manual.

Objective: To determine that the applicant:

1) Exhibits knowledge by explaining the elements of a vertical takeoff to a hover.
2) Ascends to and maintains the manufacturer's recommended hovering altitude under existing wind conditions by—
   (a) establishing and maintaining rpm within normal limits.
   (b) keeping forward and sideward movement to a minimum with no aft movement.
   (c) maintaining stationary position over a surface reference point.
   (d) maintaining desired heading, plus or minus 15°.
3) Checks engine instruments and flight controls in hover to ensure intended flight profile.

b. Task: Hover Taxi

References: AC 61-13, Helicopter Flight Manual.

Objective: To determine that the applicant:

1) Exhibits knowledge by explaining the elements of pattern flying at hovering altitude.
2) Hover taxies around a square, rectangle, or other ground reference, demonstrating forward, sideward, and rearward hovering and hovering turns.
3) Maintains rpm within normal limits.
4) Maintains desired ground track, plus or minus 3 feet, on straight legs.
5) Maintains constant rate of turn at pivot points.
6) Turns to desired headings and maintains those headings, plus or minus 15°.
7) Maintains position within 3 feet of each pivot point during turns.
8) Makes 90°, 180°, and 360° pivoting turns, stopping within 15° of desired heading.
9) Maintains the recommended hovering altitude, plus or minus 2 feet.

c. Task: Vertical Landing

References: AC 61-13, Helicopter Flight Manual.

Objective: To determine that the applicant:

1) Exhibits knowledge by explaining the elements of a vertical landing from a hover.
2) Descends from the manufacturer's recommended hovering altitude to a landing during headwind, tailwind, and crosswind conditions by—
   (a) maintaining rpm within normal limits.
   (b) establishing and maintaining a constant rate of descent.
   (c) landing within 2 feet of the designated touchdown point, with minimum forward and sideward movement, and no aft movement.
   (d) maintaining desired heading, plus or minus 15°.
3) Lowers collective pitch to the full down position after the landing gear is firmly on the surface.

d. Task: Surface Taxi

References: AC 61-13, Helicopter Flight Manual.
NOTE: Considering equipment, the examiner may test this task orally.

Objective: To determine that the applicant:

1) Exhibits knowledge by explaining safe surface taxi procedures.
2) Taxies the helicopter from one point to another under headwind, crosswind, and downwind conditions, with the landing gear in contact with the surface by—
   (a) establishing and maintaining rpm within normal limits.
   (b) maintaining speed no greater than a brisk walk.
   (c) maintaining desired ground track, plus or minus 2 feet.
   (d) maintaining desired headings, plus or minus 10°.

e. Task: Air Taxi

References: AIM.

Objective: To determine that the applicant:

1) Exhibits knowledge by explaining the elements of air taxi procedures.
2) Air taxies from one point to another.
3) Maintains rpm within normal limits.
4) Selects a safe airspeed/altitude.

4. AIRPORT, HELIPORT, AND TRAFFIC PATTERN OPERATIONS

a. Task: Airport and Heliport Marking and Lighting

References: AC 61-21, AIM.

Objective: To determine that the applicant:

1) Exhibits knowledge by explaining airport and heliport markings and lighting aids.
2) Identifies and interprets airport, heliport, taxiway marking, and lighting aids.

b. Task: Radio Communications and ATC Light Signals

References: AC 61-21, AC 61-23, AIM.

Objective: To determine that the applicant:

1) Exhibits knowledge by explaining radio communications and ATC light signals at controlled and uncontrolled airports and heliports, and prescribed procedures for radio failure.
2) Selects appropriate frequencies for the facilities to be used.
3) Transmits requests and reports using recommended standard phraseology.
4) Receives, acknowledges, and complies with radio communications.

c. Task: Normal and Crosswind Departure

References: AC 61-13, Helicopter Flight Manual.

Objective: To determine that the applicant:

1) Exhibits knowledge by explaining the elements of normal and crosswind takeoffs from a hover.
2) Maintains rpm within normal limits.
3) Accelerates to the normal climb airspeed and established proper climb power setting.
4) Maintains the desired heading plus or minus 15°.
5) Corrects for crosswind during the takeoff leg.
   NOTE: If a crosswind condition does not exist, the applicant's knowledge of the task will be evaluated through oral testing.

d. Task: Traffic Pattern Operation

References: FAR Part 91, AC 61-21, AC 61-23, AIM, Helicopter Flight Manual.

Objective: To determine that the applicant:

1) Exhibits knowledge by explaining traffic pattern procedures at controlled and uncontrolled airports and heliports, including collision, wind-shear, and wake turbulence avoidance.
2) Follows established traffic pattern procedures according to instructions or rules.
3) Corrects for wind drift to follow appropriate ground track.

4) Maintains proper spacing from other traffic.

5) Maintains rpm within normal limits.

6) Maintains traffic pattern altitude, plus or minus 100 feet.

7) Maintains desired airspeed, plus or minus 15 knots.

8) Maintains desired heading, plus or minus 15°.

9) Completes prelanding cockpit check at the proper position in the traffic pattern.

10) Maintains orientation throughout the traffic pattern.

e. Task: Normal and Crosswind Approach

References: AC 61-13, Helicopter Flight Manual.

Objective: To determine that the applicant:

1) Exhibits knowledge by explaining the elements of normal and crosswind approaches to a hover.

2) Maintains rpm within normal limits.

3) Establishes a descent at the recommended airspeed and approach angle.

4) Maintains the proper approach angle and rate of closure to the point of transition to a hover.

5) Makes a smooth transition to a hover (recommended hovering altitude, plus or minus 2 feet).

6) Terminates the approach within 3 feet of the designated point.

7) Corrects for crosswind and maintains a straight ground track.
   NOTE: If a crosswind condition does not exist, the applicant's knowledge of the task will be evaluated through oral testing.

f. Task: Go-Around

References: AC 61-13, AC 61-21, Helicopter Flight Manual.

Objective: To determine that the applicant:

1) Exhibits knowledge by explaining the elements of a go-around procedure, including making timely decisions, correct action, and coping with low rpm and yaw,

2) Makes a go-around, when it becomes necessary.

3) Applies necessary power and established the proper aircraft attitude to attain recommended climb airspeed.

4) Trims the helicopter and climbs at recommended airspeed, plus or minus 10 knots, and maintains the desired track.

g. Task: Night Flight Operations

References: AC 61-21, AC 67-2, AIM.

NOTE: Night flight operations will be evaluated only if the applicant meets night flying regulatory requirements. If the applicant is evaluated on night flight operations, the evaluation must include elements 1 through 3. The evaluation may include, at the option of the examiner, elements 4 and 8. If night flight operations are not evaluated, the applicant's certificate will bear the limitation, "Night Flying Prohibited."

Objective: To determine that the applicant:

1) Explains preparation, equipment, and factors essential to night flight.
2) Determines helicopter, heliport, airport, and navigation lighting.
3) Exhibits knowledge by explaining night flying procedures, including safety precautions and emergency actions.
4) Inspects the helicopter by following the checklist which includes items essential for night flight operations.
5) Starts, taxies, and performs pretakeoff check adhering to good operating practices.
6) Performs departures with emphasis on visual reference.
7) Navigates and maintains orientation under VFR conditions.
8) Approaches and land adhering to good operating practices for night flight operations.

5. CROSS-COUNTRY FLIGHT OPERATIONS

a. Task: Pilotage and Dead Reckoning

References: AC 61-21, AC 61-23.

Objective: To determine that the applicant:

1) Exhibits knowledge by explaining the pilotage and dead reckoning techniques and procedures.
2) Follows the preplanned course by visual reference to landmarks.
3) Identifies landmarks by relating the surface features to chart symbols.
4) Navigates by means of precomputed headings, groundspeed, and elapsed time.
5) Combines pilotage and dead reckoning.
6) Verifies helicopter position within 3 nautical miles of the flight planned route at all times.
7) Arrives at the en route checkpoints and destinations, plus or minus 5 minutes of the initial or revised ETA.
8) Estimates remaining fuel.
9) Maintains the selected altitudes, plus or minus 200 feet.
10) Maintains the desired airspeed, plus or minus 15 knots, and heading plus or minus 10°.

b. Task: Radio Navigation

References: AC 61-21, AC 61-23.

NOTE: Applicant will not be required to perform radio navigation procedures if competency has been previously shown on an FAA practical test. If the helicopter flown does not have radio navigation equipment, competence will be determined by oral testing.

Objective: To determine that the applicant:

1) Exhibits knowledge by explaining radio navigation, equipment, procedures, and limitations.
2) Selects and identifies the desired radio facility.
3) Locates position relative to the radio navigation facility.
4) Intercepts and tracks a given radial or bearing.
5) Locates position using slope radials or bearings.
6) Recognizes or describes the indication of station passage.
7) Recognizes signal loss and takes appropriate action.
8) Maintains the appropriate altitude, plus or minus 200 feet, and desired airspeed, plus or minus 15 knots.

c. Task: Diversion

References: AC 61-21.

Objective: To determine that the applicant:

1) Exhibits knowledge by explaining the procedures for diverting, including the recognition of adverse weather conditions.
2) Selects the appropriate alternate.
3) Diverts toward the alternate promptly.
4) Makes a reasonable estimate of heading, groundspeed, arrival time, and fuel consumption.
5) Maintains the appropriate altitude, plus or minus 200 feet, and desired airspeed, plus or minus 15 knots.

d. Task: Lost Procedures

References: AC 61-21.

Objective: To determine that the applicant:

1) Exhibits knowledge by explaining lost procedures, including reasons for—
    (a) maintaining the original or an appropriate heading, identifying landmarks, and climbing, if necessary.
    (b) proceeding to and identifying the nearest concentration of prominent landmarks.
    (c) using available radio navigation aids or contacting an appropriate facility for assistance.
    (d) planning a precautionary landing if deteriorating visibility and/or fuel exhaustion is imminent.

2) Selects the best course of action when given a lost situation.

6. FLIGHT MANEUVERS

a. Task: Running Takeoff

References: AC 61-13, Helicopter Flight Manual.
NOTE: This task applies only to helicopters that are equipped with wheel-type landing gear.

Objective: To determine that the applicant:

1) Exhibits knowledge by explaining the elements of running takeoffs, including situations requiring this procedure.
2) Maintains rpm within normal limits.
3) Uses a predetermined power setting below that required to hover.
4) Initiates forward accelerating movement on the surface, maintaining a straight ground run and heading, plus or minus 15°.
5) Transitions to flight with little or no pitching.
6) Climbs to and maintains an altitude not to exceed 10 feet AGL while accelerating to climb airspeed.
7) Climbs to approximately 50 feet AGL, then adjusts power to recommended climb setting.
8) Corrects for crosswind to maintain straight ground track.

b. Task: Running Landing

References: AC 61-13, Helicopter Flight Manual.
NOTE: This task applies only to helicopters that are equipped with wheel-type landing gear.

Objective: To determine that the applicant:

1) Exhibits knowledge by explaining the elements of shallow approaches and running landings, including situations requiring this procedure.
2) Maintains rpm within normal limits.
3) Established descent at the recommended airspeed and proper approach angle.
4) Maintains the proper approach angle, rate of closure, and recommended airspeed to a point approximately 50 feet AGL.
5) Makes a smooth transition from descent to surface contact while still in translational lift, using less than hovering power.
6) Contact the surface in a level attitude, beyond and within 50 feet of a designated spot.
7) Corrects properly for crosswind, maintaining the landing gear parallel with the ground track throughout the ground run.

c. Task: Slope Operation

References: AC 61-13, Helicopter Flight Manual.

NOTE: This task applies only to helicopters that are equipped with skid-type landing gear.

Objective: To determine that the applicant:

1) Exhibits knowledge by explaining the elements of slope operations.
2) Selects a suitable slope, considering wind effect and obstructions.
3) Hover taxies slowly to the selected slope, heads into the wind (headwind component), positions to land cross-slope, and avoids turning the tail rotor upslope while positioning.
4) Makes a smooth positive descent to touch the upslope skid on the sloping surface.
5) Maintains a stabilized level attitude momentarily before lowering the downslope skid to touchdown.
6) Recognizes when the slope is too steep and abandons the landing proper to using full lateral cyclic control.
7) Avoids abrupt, erratic, or overcontrolling with cyclic, collective, or anti-torque pedals.
8) Makes a smooth transition from the slope to a stabilized attitude momentarily prior to lifting off vertically to a hover.
9) Moves slowly away from the slope and avoids turning the tail upslope.
10) Maintains desired heading, plus or minus 5°, during slope landings and takeoffs.
11) Maintains rpm within normal limits.

d. Task: Rapid Deceleration

References: AC 61-13, Helicopter Flight Manual.

Objective: To determine that the applicant:

1) Exhibits knowledge by explaining the elements of rapid deceleration.
2) Initiates maneuver properly.
3) Decelerates and terminates in a stationary hover at recommended hovering altitude.
4) Maintains heading, plus or minus 15°.
5) Maintains rpm within normal limits.

e. Task: Confined Area and Pinnacle Operations

References: AC 61-13.

NOTE: This task should be evaluated using unimproved training sites that offer little hazard with respect to obstacles.

Objective: To determine that the applicant:

1) Exhibits knowledge of confined area and pinnacle operations.
2) Performs a high and low reconnaissance to—
   (a) evaluate wind, terrain, and obstructions.
   (b) select a proper approach path, touchdown point, and departure path.
3) Tracks the selected approach path toward the touchdown point at an acceptable approach angle.
4) Performs a low reconnaissance during the approach to verify findings on the high reconnaissance.
5) Terminates the approach, either to a hover or to the surface, as appropriate, at the selected touchdown point.
6) Performs a ground reconnaissance to—
   (a) evaluate wind and obstructions.
   (b) select a proper takeoff point.
   (c) plan a safe hover taxi.
   (d) place ground markers, if required.
7) Performs a proper takeoff, safely clears obstructions, and tracks the preselected departure path.
8) Maintains rpm within normal limits.

7. EMERGENCY OPERATIONS

a. Task: Power Failure at a Hover

References: AC 61-13, Helicopter Flight Manual.

Objective: To determine that the applicant:

1) Exhibits knowledge by explaining the procedures to use when power failure occurs at a hover.
2) Establishes either a stationary or forward hover into the wind at recommended rpm.
3) Performs hovering autorotation when examiner simulates power failure.
4) Touches down with acceptable forward movement, minimum sideward movement, no aft movement, and without excessive loads on the landing gear.
5) Maintains heading, plus or minus 10°.

b. Task: Power Failure at Altitude

References: AC 61-13, Helicopter Flight Manual.
NOTE: No simulated power failure at altitude will be given where an actual touchdown could not be safely completed if it should become necessary, nor where an autorotative descent might constitute a violation of FARs. The examiner will direct the applicant to terminate this task in a power recovery at an

altitude high enough to assure that a safe touchdown could be accomplished in the event of an actual power failure.

Objective: To determine that the applicant:

1) Exhibits knowledge by explaining autorotation entry, maneuvering, and landing procedures following power failure at altitude.
2) Enters autorotation promptly when an examiner simulates power failure by—
   (a) lowering the collective pitch, as necessary, to maintain rotor rpm within acceptable limits, and
   (b) establishing the recommended autorotation airspeed.
3) Flies autorotation pattern by—
   (a) selecting a suitable landing area.
   (b) establishing an autorotation pattern appropriate for position, attitude, and wind conditions.
   (c) maintaining rotor rpm within normal limits.
   (d) maintaining proper pedal trim.
   (e) arriving at the selected area with proper altitude, landing attitude, and acceptable rotor rpm and groundspeed.
4) Terminates the autorotation as directed by the examiner by performing a proper power recovery.

c. Task: System and Equipment Malfunctions

References: AC 61-13, Helicopter Flight Manual.

Objective: To determine that the applicant:

1) Exhibits knowledge by explaining causes of, indications of, and pilot actions for malfunctions of various systems and equipment.
2) Takes appropriate action for simulated malfunctions such as—
   (a) carburetor or induction icing, if appropriate.
   (b) fuel starvation.
   (c) smoke in cockpit.
   (d) engine compartment fire.
   (e) electrical system malfunction.
   (f) hydraulic system malfunction, if appropriate.
   (g) trim inoperative, if appropriate.
   (h) other malfunctions outlined in the helicopter flight manual.

d. Task: Antitorque System Failure

References: AC 61-13, Helicopter Flight Manual.
NOTE: At the discretion of the examiner, this task may be orally tested.

Objective: To determine that the applicant:

1) Exhibits adequate knowledge by explaining antitorque system failure procedures while hovering and at altitude.
2) Recognizes simulated antitorque failure in a hover or in cruising flight and takes immediate and proper action.

e. Task: Settling-With-Power

References: AC 61-13, Helicopter Flight Manual.
NOTE: At the discretion of the examiner, this task may be orally tested.

Objective: To determine that the applicant:

1) Exhibits knowledge by explaining the conditions which cause settling-with-power and procedures used to recover to normal flight.
2) Demonstrates settling-with-power entry.
3) Recovers immediately and correctly at first indications of settling-with-power.

f. Task: Low Rotor RPM Recovery

References: AC 61-13, Helicopter Flight Manual.
NOTE: At the discretion of the examiner, this task may be orally tested.

Objective: To determine that the applicant:

1) Exhibits knowledge by explaining the conditions which cause low rotor rpm and the procedure used to recover.
2) Recognizes low rotor rpm and takes immediate and proper recovery action.

g. Task: Dynamic Rollover

References: Helicopter Flight Manual.
NOTE: This task will be orally tested only.

Objective: To determine that the applicant:

1) Exhibits knowledge by explaining helicopter lateral rolling tendencies during certain ground operations.
2) Understands the interaction between tail rotor thrust, crosswind, slope, center of gravity, and cyclic and collective pitch control in contributing to dynamic rollover.
3) Demonstrates preventive flight technique during takeoffs, landings, and slope operations.

h. Task: Ground Resonance

References: AC 61-13, Helicopter Flight Manual.
NOTE: This task will be orally tested only.

Objective: To determine that the applicant:

1) Exhibits knowledge by explaining the hazards associated with ground resonance.
2) Understands the conditions that contribute to ground resonance.
3) Demonstrates preventive flight technique during takeoffs, landing, and slope operations, if appropriate.

NOTE: References used in PRIVATE PILOT PRACTICAL TEST STANDARDS

FAR Part 61	Certification: Pilots and Flight Instructors
FAR Part 91	General Operating and Flight Rules
AC 00-6	Aviation Weather
AC 00-45	Aviation Weather Sources
AC 61-13	Basic Helicopter Handbook
AC 61-12	Flight Training Handbook
AC 61-23	Pilot's Handbook of Aeronautical Knowledge
AC 61-27	Instrument Flying Handbook
AC 61-84	Role of Preflight Preparation
AC 67-2	Medical Handbook for Pilots
AC 91-13	Cold Weather Operation of Aircraft
AC 91-55	Reduction of Electrical Systems Failure Following Engine Starting
AIM	Airman's Information Manual
	Helicopter Flight Manual for helicopter being used for the practical test

Private Pilot—Helicopter Practical Test Checklist

1. PREFLIGHT PREPARATION
   - ☐ a. Certificates and Documents
   - ☐ b. Obtaining Weather Information
   - ☐ c. Determining Performance and Limitations
   - ☐ d. Cross-Country Flight Planning
   - ☐ e. Operation of Helicopter Systems
   - ☐ f. Aeromedical Factors
2. GROUND OPERATIONS
   - ☐ a. Visual Inspection
   - ☐ b. Cockpit Management

   ☐ c. Starting Engine
   ☐ d. Pretakeoff Check
   ☐ e. Postflight Procedures

3. HOVERING AND MANEUVERING BY GROUND REFERENCES
   ☐ a. Vertical Takeoff
   ☐ b. Hover Taxi
   ☐ c. Vertical Landing
   ☐ d. Surface Taxi
   ☐ e. Air Taxi

4. AIRPORT, HELIPORT, AND TRAFFIC PATTERN OPERATIONS
   ☐ a. Airport and Heliport Marking and Lighting
   ☐ b. Radio Communications and ATC Light Signals
   ☐ c. Normal and Crosswind Departure
   ☐ d. Traffic Pattern Operation
   ☐ e. Normal and Crosswind Approach
   ☐ f. Go-Around
   ☐ g. Night Flight Operations

5. CROSS-COUNTRY FLIGHT OPERATIONS
   ☐ a. Pilotage and Dead Reckoning
   ☐ b. Radio Navigation
   ☐ c. Diversion
   ☐ d. Lost Procedures

6. FLIGHT MANEUVERS
   ☐ a. Running Takeoff
   ☐ b. Running Landing
   ☐ c. Slope Operation
   ☐ d. Rapid Deceleration
   ☐ e. Confined Area and Pinnacle Operations

7. EMERGENCY OPERATIONS
   ☐ a. Power Failure at a Hover
   ☐ b. Power Failure at Altitude
   ☐ c. System and Equipment Malfunctions
   ☐ d. Antitorque System Failure
   ☐ e. Settling-with-Power
   ☐ f. Low Rotor RPM Recovery
   ☐ g. Dynamic Rollover
   ☐ h. Ground Resonance

Appointment with Inspector or Examiner:

Name_____

Time/Date_____

Acceptable Aircraft

☐ Aircraft Documents: Airworthiness Certificate, Registration
☐ Certificate, Operating Limitations
☐ Aircraft Maintenance Records: Airworthiness Inspections
☐ FCC Station License

Personal Equipment

☐ Current Aeronautical Charts
☐ Computer and Plotter
☐ Flight Plan Form
☐ Flight Logs
☐ Current AIM

Personal Records

☐ Pilot Certificate
☐ Medical Certificate
☐ Completed FAA Form 8710-1, Airman Certificate and/or Rating
☐ Application
☐ AC Form 8080-2, Airman Written Test Report
☐ Logbook with Instructor's Endorsement
☐ Notice of Disapproval (if applicable)
☐ Approved School Graduation Certificate (if applicable)
☐ FCC Radiotelephone Operator Permit (if applicable)
☐ Examiner's Fee (if applicable)

Appendix C
Torque Experiment

MY FATHER, WHO IS AN ENGINEER, THOUGHT OF THIS EXPERIMENT while he was proofreading the manuscript. The experiment is a good demonstration of Newton's Third Law (for every action there's an equal and opposite reaction) and will help you understand why a helicopter needs some way to counteract torque (FIG. C-1).

Materials:

- Soda straw, shortened if necessary.
- Rubber band that is longer than the straw.
- Two pencils

Directions

1. Feed the rubber band through the straw so that loops stick out each end of the straw.

2. Insert a pencil through each loop in the rubber band.

3. Have another person hold the straw gently at its midpoint. Do not squeeze the straw.

4. Hold one pencil steady while turning the other pencil to wind the rubber band.

5. When the rubber band is wound moderately tight, release both pencils. The assistant should keep holding the straw at the midpoint.

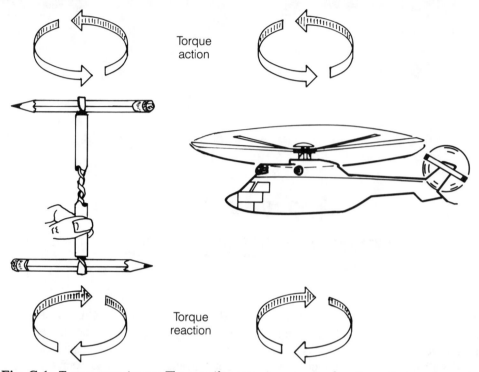

Fig. C-1. *Torque experiment: The pencils rotate in opposite directions, just as the main rotor and fuselage of a helicopter would rotate in opposite directions if not counteracted by the tail rotor.*

Observations

Both pencils should spin, but in opposite directions, caused by the torque force stored in the twisted rubber band. The rubber band is like the engine in the helicopter. One pencil is the rotor and the other is the fuselage. When the rotor is turned by the engine, the fuselage of the helicopter wants to spin in the opposite direction because of the engine torque; therefore, a helicopter must have some device, in conventional helicopters it's the tail rotor, that provides a force to counteract the engine torque that is transmitted to the main rotor.

Appendix D
Watch Out for Power Lines

TOWERS, POLES, AND THEIR ELECTRICAL LINES MIGHT NOT ALWAYS BE readily visible. Most are not required to be marked under the FAA criteria that determine what is considered to be an obstruction to air navigation. Also, under some conditions, such as sun glare or haze, it can be difficult to see the lines running between the support structures. Taking two simple steps can greatly reduce the chances of accidentally contacting electrical facilities while you're flying.

First, take time for safety prior to the flight. Check the aeronautical charts for obstructions when you plan the route. Certain electrical lines are charted because of their height. Also, airport directories carry warnings of power lines located close to runways.

Then, observe the minimum altitude requirements while airborne, especially the 1,000-foot minimum over populated areas. It's a good idea to stay above 1,000 feet when flying over lakes, rivers, or canyons to avoid any power line crossings.

This step-by-step list is an excellent summation:

1. Before you take off, check the airport directory for warnings about power lines at your destination.

2. Check your route and be familiar with marked obstructions.

3. Always observe altitude minima.

4. Do not allow adverse weather conditions to force you to fly too close to the ground. Check aviation weather forecasts to make sure cloud heights provide an adequate ceiling for safe visual flying.

Fig. D-1. *Bell 212 pulling lead-line over transmission towers.*

5. Remember that sun glare can make power lines nearly invisible.

6. Maintain a safe altitude over rivers, lakes, and other waterways.

7. When visibility is poor, increase your altitude above minima or fly instrument flight rules (IFR).

8. Be aware that power lines and towers are marked only near airports and at certain water crossings.

9. When you're using private airports, check for nearby power lines. Call ahead for information about any obstruction at any private field you're planning to use.

10. When flying through a gap or over a mountain ridge, watch for winds and turbulence that could force you into a power line crossing.

(List is courtesy of the Pennsylvania Power & Light Company in conjunction with the FAA.)

Appendix E
Commonly Used
Hand Signals

Helicopter operations around the world commonly understand a prescribed set of hand signals. Ground personnel working near flying helicopters should wear safety glasses, hard hats, and hearing protection.

Fig. E-1. *Sikorsky SH-60B SeaHawk landing on the U.S.S. McInerney.*

United Technologies Sikorsky Aircraft

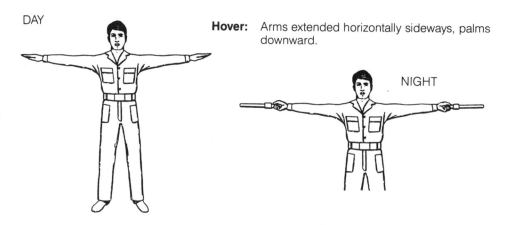

DAY

Hover: Arms extended horizontally sideways, palms downward.

NIGHT

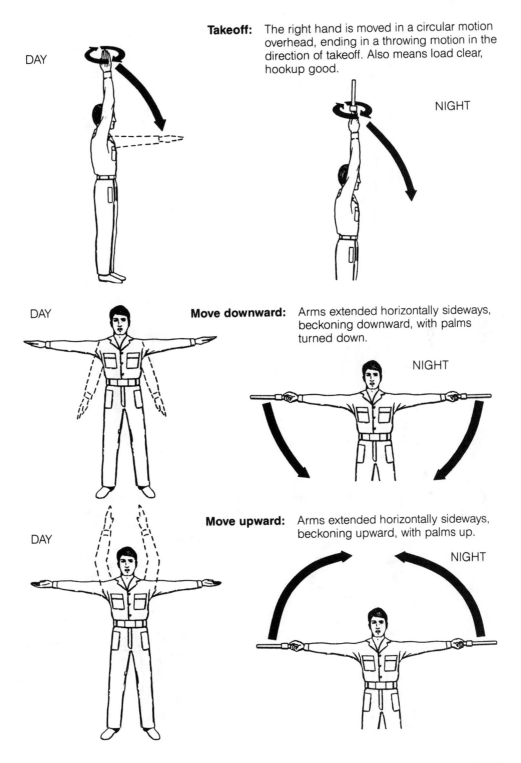

Takeoff: The right hand is moved in a circular motion overhead, ending in a throwing motion in the direction of takeoff. Also means load clear, hookup good.

DAY

NIGHT

Move downward: Arms extended horizontally sideways, beckoning downward, with palms turned down.

DAY

NIGHT

Move upward: Arms extended horizontally sideways, beckoning upward, with palms up.

DAY

NIGHT

335

DAY

Move forward: Arms a little aside, palms facing backward and repeatedly moved upward-backward from shoulder height.

NIGHT

DAY

Move rearward: Arms by sides, palms facing forward, arms swept forward and upward repeatedly to shoulder height.

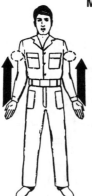

NIGHT

Move to left: Right arm extended horizontally sideways in direction of movement and other arm swung in front of body in same direction, in a repeating movement.

DAY

NIGHT

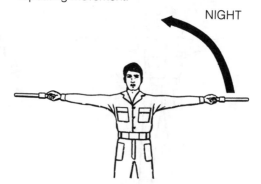

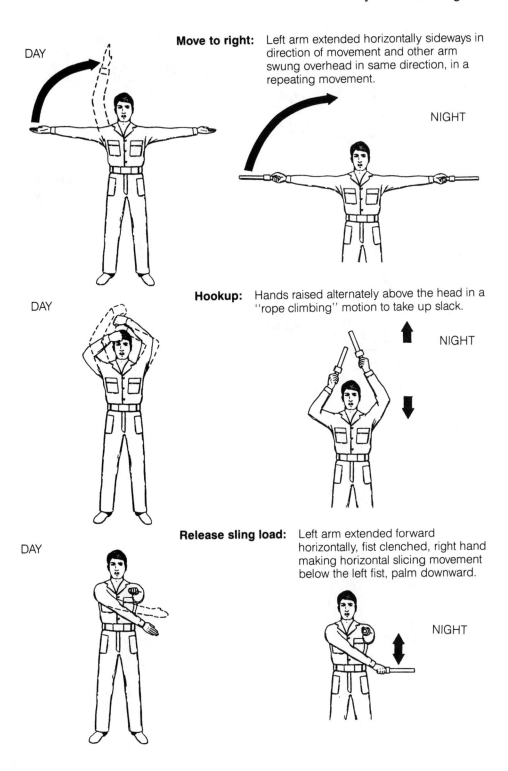

DAY

Move to right: Left arm extended horizontally sideways in direction of movement and other arm swung overhead in same direction, in a repeating movement.

NIGHT

DAY

Hookup: Hands raised alternately above the head in a "rope climbing" motion to take up slack.

NIGHT

DAY

Release sling load: Left arm extended forward horizontally, fist clenched, right hand making horizontal slicing movement below the left fist, palm downward.

NIGHT

DAY

Move hook down or up: Right fist held above head: left arm extended horizontally, palm faced outward, then swept down or up to indicate direction of hook movement.

NIGHT

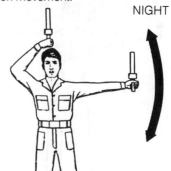

DAY

Negative signal: Hand raised, thumb down.

NIGHT

DAY

Affirmative signal: Hand raised, thumb up.

NIGHT

338

DAY   **Stop:**   Arms held crossed overhead.

NIGHT

Land: Arms crossed and extended downward in front of the body.

DAY

NIGHT

DAY   **Cut engine(s):**   Either arm and hand level with shoulder, hand moving across throat.

NIGHT

Glossary

advancing blade—That half of the rotor disc in which rotation of the blade is moving in the same direction as the movement of the helicopter. If a helicopter with main rotor blades that rotate counterclockwise is moving forward, the advancing blade will be in the right half of the rotor disc; if moving backward, the advancing blade will be in the left half; if moving sideward to the left, the advancing blade will be in the forward half; and if moving sideward to the right, the advancing blade will be in the rear half.

AFCS (automatic flight control system)—An electromechanical system installed to improve the aerodynamic stability of an aircraft. An AFCS cannot program the aircraft's flight as does a conventional autopilot. It only serves to keep the aircraft in the flight condition established by the pilot.

airfoil—Any surface designed to obtain a useful lift reaction when moving through the air at a specified velocity.

angle of attack—The acute angle between the chordline of an airfoil and the relative wind.

articulated rotor—A rotor system in which the blades are free to flap, drag, and feather.

ATIS (automatic terminal information service)—A recorded announcement provided at busy airports that is broadcast on a discreet radio frequency and contains information about current weather, active runways, flight hazards, and other information necessary for safe operations.

autorotation—The process of producing lift with freely-rotating airfoils by means of the aerodynamic forces resulting from an upward flow of air.

axis of rotation—An imaginary line that passes through a point about which a body rotates and is perpendicular to the plane of rotation.

BIM (blade inspection method)—A means of checking the integrity of main rotor blades. The blades are sealed and pressurized with an inert gas. If a crack develops in the blade, the gas leaks out, the pressure goes down, and the BIM indicator changes color.

blade damper—A device (spring, friction, or hydraulic) installed on the vertical (drag) hinge to diminish or dampen blade oscillation (hunting) around this hinge.

blade flapping—The movement of the rotor blades in a vertical plane (up and down) due to their attachment to the rotor hub by a horizontal hinge.

blade loading—The load placed on the rotor blades of a helicopter, determined by dividing the gross weight of the helicopter by the combined area of all the rotor blades.

center of gravity—An imaginary point where the resultant of all weight forces in a body can be considered to be concentrated for a position of the body.

center of pressure—The imaginary point on the chord line of an airfoil where the resultant of all aerodynamic forces can be assumed to act.

centrifugal force—The force created by the tendency of a body to follow a straight-line path against the force that causes it to move in a curve, resulting in a force that tends to pull away from the axis of rotation.

chip light—Warning light that illuminates to inform the pilot that a metallic chip has been detected in a particular component.

chord—An imaginary straight line between the leading and trailing edges of an airfoil.

collective pitch control (or lever)—The control by which the pilot can vary the pitch of all the rotor blades equally and simultaneously.

confined area—An area where the flight of the helicopter is limited in some direction by terrain or the presence of obstructions.

coning—The upward bending of the rotor blades caused by the combined forces of lift and centrifugal force.

conservation of angular momentum, law of—As the center mass of a rotating body moves closer to the axis of rotation, the velocity of the body increases.

coriolis effect—The tendency of a mass to increase or decrease its angular velocity when its radius of rotation is shortened or lengthened, respectively. In a helicopter, the coriolis effect of the main rotor blades is compensated for by the lead-lag hinges.

correlation device—A mechanical linkage between the collective pitch lever and the throttle designed so that the throttle opens as the collective lever is raised and closes as the collective lever is lowered. The correlation device reduces some of the burden of adjusting the throttle to maintain rotor rpm whenever the pilot makes changes in collective pitch.

cyclic pitch control or stick—The control by which the pilot changes the pitch of the rotor blades individually during a cycle of revolution to control the tilt of the rotor disc, which affects the direction and velocity of horizontal flight.

dead man's curve—*See* height-velocity diagram definition.

density altitude—Pressure altitude corrected for temperature and humidity.

disc area—The area swept by the blades of the rotor. This is a circle with its center at the hub axis and a radius of longest blade length.

disc loading—The ratio of helicopter gross weight to rotor disc area.

dissymmetry of lift—The unequal lift across the rotor disc area resulting from the difference in the velocity of air over the advancing blade half and the retreating blade half of the rotor disc area.

DME (distance measuring equipment)—A navigation aid that provides a pilot with the helicopter's straight-line distance in nautical miles from the selected ground station.

drag—That component of the resultant of the aerodynamic forces acting on a body that acts parallel to the relative airflow; the force that tends to resist movement of an airfoil through the air.

drag hinge—*See* vertical hinge.

dual tachometer or tach—A single instrument with two indications displayed: engine revolutions per minute (rpm) and main rotor rpm of a single-engine helicopter. A triple tach displays respective engine rpm of a twin-engine helicopter plus main rotor rpm.

EFIS (electronic flight instrument system)—Devices designed to replace electromechanical gauges (airspeed indicator, altimeter, compass, navigation instruments, engine instruments, and the like) with video display screens that provide virtually unlimited possibilities for presenting flight data. *See also* glass cockpit.

FAA—Federal Aviation Administration.

feathering axis—The axis about which the pitch angle of a rotor blade is varied. Sometimes referred to as the spanwise axis.

feathering action—Action that changes the pitch angle of the rotor blades periodically by rotating the blades around their feathering axis.

flapping—The vertical movement of a blade about a flapping hinge.

flapping hinge—The hinge with its axis parallel to the rotor plane of rotation, which permits the rotor blades to flap to equalize lift between the advancing blade half and the retreating blade half of the rotor disc.

force—Any influence that changes or tends to change the state of rest or uniform motion of a body.

gizmo—Any device that adds complication to simplify anything complicated.

glass cockpit—An aircraft cockpit dominated by electronic flight instrument systems (EFIS).

ground effect—The cushion of denser air confined beneath the rotor system of a hovering helicopter that gives additional lift and thus decreases the power required to hover.

ground resonance—The destructive combination of rotor blade lead-lag motion with the aircraft rocking on its landing gear.

gyroplane—A rotorcraft that derives the whole or a substantial part of its lift from a freely-rotating rotor.

gyroscopic precession—A characteristic of all rotating bodies. When a force is applied to the periphery of a rotating body parallel to its axis of rotation, the rotating body will tilt in the direction of the applied force 90 degrees later in the plane of rotation.

height-velocity diagram—The combinations of altitude and airspeed from which structural damage to the helicopter will occur in case of a power failure; also called the dead man's curve.

helicopter—A rotorcraft deriving the whole or a substantial part of its lift from one or more power-driven rotors (from Greek words *heliko*, meaning spiral-like, and *pteron*, meaning wing).

hingeless rotor—A rotor with no mechanical hinges.

hovering in ground effect—Maintaining a fixed position over a spot on the surface that compresses a cushion of high-density air between the main rotor and the surface and thus increases the lift produced by the main rotor. Normally the main rotor must be within one-half rotor diameter to the surface in order to produce an efficient ground effect.

hovering out of ground effect—Maintaining a fixed position over a spot on the surface at some altitude above the surface at which no additional lift is obtained from ground effect.

HUMS (health and usage monitoring systems)—Sophisticated electronic and computer devices that, by continually monitoring parameters such as temperature, pressure, and vibration level, can determine the condition, or health, of mechanical components. In helicopters, the engines, main gearbox, drivetrain, and tail rotor gearbox are typical candidates for HUMS.

hunting—The tendency of a blade, due to coriolis effect, to seek a position ahead of or behind that would be determined by centrifugal force alone.

ILS (instrument landing system)—A precision approach system that provides the pilot with horizontal guidance and vertical guidance to the proper touchdown point on the runway.

Kollsman window—The window of an altimeter that indicates current barometric pressure (the altimeter setting) for a proper indication of the aircraft's elevation above sea level. A pilot adjusts the window's indication by setting the current barometric pressure as reported by an air traffic controller or specialist or a weather observer.

lead-lag hinge—*See* vertical hinge.

lift—That component of the resultant of the aerodynamic forces acting on a body that acts at right angles to the relative airflow; the force derived from an airfoil according to Bernoulli's Principle (venturi effect).

manifold pressure—The air pressure measured at a certain point in the induction manifold of a reciprocating engine.

maximum performance takeoff—Takeoff procedure used when operating from small or restricted areas.

nautical mile—One minute of arc on a meridian (one minute of latitude); one minute of arc on the earth's equator. One nautical mile equals approximately 1.15 statute miles or 1.85 kilometers; therefore, to convert knots to miles per hour simply multiply knots by 1.15; to convert knots to kilometers per hour, double knots for a rough approximation, or multiply the knots by 1.85, to be more precise.

NAVSTAR—Navigation system that utilizes satellites to determine the position of objects on the earth's surface or in the air (global positioning system (GPS)).

Newton's Third Law—To every action there is an equal and opposite reaction.

pendular action—The tendency of the fuselage of a helicopter to oscillate freely longitudinally and laterally, similar to the action of a pendulum. Pendular action can be exaggerated by overcontrolling.

pinnacle—An area from which the surface drops away steeply on all sides.

PIO (pilot induced oscillations)—Changes in the aircraft's position caused by normal human reaction delays.

PIT (pilot induced turbulence)—Changes in the aircraft's position caused by normal human reaction delays. More properly known as pilot induced oscillations (PIO).

pitch angle—The acute angle between the chord line of a rotor blade and the reference plane of the rotor hub or the plane of rotation.

plane of rotation—Also called tip-path plane, formed by the average tip-path of the rotor blades, perpendicular to the axis of rotation.

raison d'être—Reason or justification for existence.

reciprocating engine—An engine powered by the to-and-from motion of pistons in enclosed cylinders, which is transformed into circular motion by a crankshaft.

relative wind or airflow—The velocity of the air with reference to a body in it; the direction of the airflow with respect to an airfoil.

resultant and component—If two or more forces act on a rigid body, and if a single force can be found whose effect upon the body is the same as that of these forces, then the single force is the resultant and the original forces are called components of that force.

ridgeline—A long area from which the surface drops away steeply on one or two sides, such as a bluff or precipice.

rigid rotor—A rotor system with blades fixed to the hub in such a way that they can feather but cannot flap or drag.

rotor—A system of rotating airfoils.

rotor disc—The circular area defined by the sweep of the rotor blades. For some aerodynamic purposes, the rotor disc can be considered as if it were a solid wing.

rpm (revolutions per minute)—The unit of measurement of a rotating object, for instance rotor rpm or engine rpm.

seat-of-the-pants—Flying by sight, touch, and feeling, ideally with the wind whistling around your goggles and ears....

semirigid rotor—A rotor system in which the blades are fixed to the hub but are free to flap and feather.

settling with power—The condition of flight when a helicopter is descending in its own rotorwash, characterized by a high vertical rate of descent, powered flight (not autorotation), and low airspeed. More power increases the rate of descent.

slip—The controlled flight of a helicopter in a direction not in line with its fore and aft axis.

slope—Any ground that is not level.

stall—The loss of streamlined airflow caused by excessive angle of attack that results in a swirling, turbulent airflow, a sudden increase in pressure on top of the airfoil, and a large loss of lift.

standard atmosphere—Atmospheric conditions in which the air is a dry, perfect gas; the temperature at sea level is 59 °F/15 °C; the pressure at sea level is 29.92 inches/1013.2 millibars of mercury (Hg); the temperature gradient is approximately 3.5 °F/2 °C per 1,000-foot change in altitude.

swashplate assembly—Part of the main rotor system responsible for transmitting changes in cyclic and collective to the main rotor blades.

switchology—Instrument and switch design. In the context of human factors, switchology is important because switches and instruments that are poorly designed or placed can cause people to operate or interpret them incorrectly.

tachometer—A device for indicating speed of rotation, usually in revolutions per minute (rpm) or, in many helicopters, percent of normal rpm.

thrust—The forward acting force of an aircraft in flight. In a helicopter, the rotor disc produces thrust and lift. The tilt of the disc determines the relative size and direction of the thrust vector.

tip-path plane—*See* plane of rotation.

tip speed—The rotative speed of the rotor at its blade tips.

tip stall—The stall condition on the retreating blade that occurs at high forward airspeeds.

torque—A force or combination of forces that tends to produce a counterrotating motion. In a single rotor helicopter where the rotor turns counterclockwise, the fuselage tends to rotate clockwise, as seen from the cockpit looking forward.

translating drift—The tendency of the entire helicopter to move in the direction of tail rotor thrust when hovering.

translational lift—The additional lift a helicopter gains as it moves from hovering into forward flight due to the increased efficiency of the rotor system.

vector—A straight line drawn to represent the magnitude, direction, and sense of a quantity such as force or velocity.

vertical hinge—Also called a lead-lag or drag hinge, permits the blades to move back and forth independently of each other. The ability to lead and lag is necessary to relieve the stresses that build up in the blades due to the coriolis effect on the blades.

VOR (very-high-frequency omnidirectional range)—A navigation aid that generates and transmits directional information from ground equipment to the aircraft, providing 360 magnetic courses to and from the station. VOR is presently the primary navigation system for civil aviation in the United States and most other countries.

V_{ne}—Never exceed speed.

V_{no}—Normal operating speed.

V_y—Speed for the best rate of climb.

Index